ESTIMATION AND CONTROL WITH
QUANTIZED MEASUREMENTS

ESTIMATION AND CONTROL WITH QUANTIZED MEASUREMENTS

RENWICK E. CURRY

Research Monograph No. 60
THE M.I.T. PRESS
Cambridge, Massachusetts,
and London, England

I would like to thank the Institute of Electrical and Electronic Engineers for permission to reprint material from the following articles:

R. E. Curry, "A New Algorithm for Suboptimal Stochastic Control," *IEEE Trans. Automatic Control*, **AC-14**, 533–536 (October 1969);

R. E. Curry, "A Separation Theorem for Nonlinear Measurements," *IEEE Trans. Automatic Control*, **AC-14**, 561–564 (October 1969);

R. E. Curry and W. E. Vander Velde, "An Extended Criterion for Statistical Linearization," *IEEE Trans. Automatic Control* (February 1970);

R. E. Curry, W. E. Vander Velde, and J. E. Potter, "Nonlinear Estimation with Quantized Measurements: PCM, Predictive Quantization, and Data Compression," *IEEE Trans. Inform. Theory* (March 1970).

Set in Monophoto Times Roman
Printed and bound in the United States of America by
The Colonial Press Inc., Clinton, Massachusetts

ISBN 0262 03037 3 (*hardcover*)

Library of Congress catalog card number: 71-103894

To Sue
Katie
Smith
Susan

Contents

Foreword

This is the sixtieth volume in the M.I.T. Research Monograph Series published by the M.I.T. Press. The objective of this series is to contribute to the professional literature a number of significant pieces of research, larger in scope than journal articles but normally less ambitious than finished books. We believe that such studies deserve a wider circulation than can be accomplished by informal channels, and we hope that this form of publication will make them readily accessible to research organizations, libraries, and independent workers.

Howard W. Johnson

Preface

The demand on digital facilities such as communication systems and data-storage systems is constantly increasing. In the past the pressures have been relieved by upgrading their capacities; primarily through the advances being made in digital hardware, attention has turned to the problem of using these facilities more efficiently. Many of the efforts are described in a general way as "data-compression," "redundancy-reduction," and "bandwidth-compression," and most of them rely on the quantization and subsequent reconstruction of data.

This monograph presents the results of some research pertaining to the distinct but related tasks of efficient estimation and control based on quantized measurements. It has been published in the hope that both researchers and engineers will find some useful ideas to expand and adapt to their own needs. The reader is assumed to have a familiarity with probability theory and random processes (at the level of Lanning and Battin, *Random Processes in Automatic Control*, or Papoulis, *Probability, Random Variables, and Stochastic Processes*), and a basic understanding of estimation theory.

Discrete-time problems are considered and the emphasis is placed on coarsely quantized measurements and on linear and, when pertinent, time-varying systems. The heart of the material is a new interpretation and outlook on the problem of generating nonlinear estimates from quantized measurements. The development of the minimum variance, or conditional mean estimate is quite fundamental, since it lays the groundwork for other types of estimates. Approximate and more easily implemented nonlinear filters are examined in some detail, especially in conjunction with three communication systems, so that the subject matter is not limited to theory alone. The design of optimal linear estimators is re-examined, and their performance is compared with that of nonlinear filters. Not surprisingly, consideration of the control of stochastic systems that have quantized measurements leads to insights into, and control strategies for, systems with other types of nonlinearities.

The major portion, but by no means all, of this research was submitted as a Ph.D. thesis to the Department of Aeronautics and Astronautics at the Massachusetts Institute of Technology. This monograph is an extensive revision of the original work, and it incorporates new analytical and numerical results and the welcome comments of many anonymous reviewers.

I wish to thank the members of my thesis committee, Professor Wallace E. Vander Velde, chairman, Professor Arthur E. Bryson, Jr., and Professor James E. Potter for their pertinent questions and fruitful suggestions. My association with them has been extremely valuable, for not only are they outstanding teachers, but their approach to technical problems is one that I particularly admire. Mr. Charles F. Price acted as a sounding board for many ideas and provided comments and suggestions about the initial draft; his help is greatly appreciated. I should also like to acknowledge the profitable discussions with Professor Terrence Fine, Dr. Herbert Gish and Dr. Donald Fraser. Finally I would like to thank my wife, Susan, for typing the initial draft and its revisions and for her patience and encouragement.

This research was supported by Grant NGR 22-009-010 from the National Aeronautics and Space Administration, monitored by Mr. Jules I. Kanter of NASA Headquarters, Washington, D.C., and was carried out at the Measurement Systems Laboratory (formerly the Experimental Astronomy Laboratory) at M.I.T.

R. E. Curry

Ithaca, New York
June, 1969

ESTIMATION AND CONTROL WITH QUANTIZED MEASUREMENTS

1 Introduction

1.1 Background

The mathematical operation of quantization exists in many communication and control systems. Quantizing elements may appear as digital transducers, analog-to-digital converters, or digital-to-analog converters; even the digital computer itself is a quantizing device, because of its finite word length. Measurements will be quantized if they are produced by digital sensors, transmitted over a digital communication link, or processed by a digital computer.

Quantization of measurements is the irreversible process of rounding arithmetical numbers, and information is lost by this operation. The word *quantizer* is reminiscent of a nonlinear input-output staircase graph. Although this is a valid representation in many instances, it is overly restrictive, and we shall interpret the information in a quantizer's output in a different manner.

A quantizer is any zero-memory input-output device that designates the interval or intervals in which its input lies.

This definition enlarges the class of nonlinear devices that may be considered quantizers. For now we see that elements containing dead-zone and saturation regions may be viewed as quantizers. Thus, regardless of the input-output graph, a quantized measurement will mean that the measured quantity lies in a known region.

The study of coarsely quantized measurements is important at this time because in many cases it is less expensive to add equipment for a more complicated processing of coarsely quantized data than to expend time, money, and effort on reducing the size of the quantum intervals. This possibility is a direct result of the advances and improvements that are being made in digital hardware.

For instance, the ever-increasing demand to transmit and store more information is being felt in almost all areas. The usual approach has been to upgrade the transmission and storage capacity of a system, but now it appears that further increases in performance will be attained most economically by a more efficient use of existing digital facilities (*IEEE*, 1967). This can be done by representing the same amount of information with fewer bits, but more sophisticated data-processing is required to recover the information.

For an example of this approach suppose that the word length in a digital telemetry system could be shortened and that the data-processing could be so employed as to keep the information loss at an acceptable value. Then the designers of the system would have the following tradeoffs, or a combination of them, at their disposal:

1. reduce the transmitter's power, size, and weight and send the same information in the same amount of time;
2. use the same transmitter to send the same information in a shorter time; this would allow more instruments to be installed and monitored.

Coarsely quantized measurements occur in other situations, as in the taking of sonar bearings (Korsak, 1967) or during the alignment of some inertial platforms. In the latter, as the platform approaches a horizontal position, the horizontal component of the specific force of the platform support becomes very small. Its detection by pulse-rebalanced accelerometers, which provide quantized information about velocity, is a lengthy, if not difficult, problem: the pulses occur at a frequency that is proportional to the platform deviation (which is small) and inversely proportional to the velocity increment of each pulse (which may be large).

One solution is to change the velocity increment of each pulse, but this may have an adverse effect on the accelerometer's accuracy in other modes of operation. An alternative approach during the alignment is to use the information that a pulse has *not* occurred to improve the knowledge of the platform angle. This is equivalent to a quantized measurement.

The example of platform alignment leads to the question of control with quantized measurements. Control and estimation are interrelated through their connection with the system as a whole, and the measurement device (in this case a quantizer) may be expected to influence the choice of control laws that give satisfactory results. In some cases, especially in those of optimal stochastic control, the interdependence of estimation and control is so strong that the two functions merge and must be designed as a single entity.

1.2 Objectives and Scope

The objective of this work is to examine the two distinct but related problems of optimal estimation and control with arbitrarily quantized measurements. Consideration is limited to discrete-time problems, and emphasis is placed on coarsely quantized measurements and linear, possibly time-varying, systems. A quadratic criterion is used in the optimal control analyses.

1.3 Estimation with Quantized Measurements

Wiener (1966) was among the first to deal with the optimal estimation of stochastic processes. He derived an integral equation for the weighting function of the optimal, linear, realizable filter for minimizing a mean-square-error criterion and solved the integral equation by spectral factorization. Later attempts (see Davenport, 1958) were made to remove such restrictions as the need for an infinite amount of data. Both Kalman (1960) and Swerling (1959), using the state-space approach, showed that the solution to the problem of optimal linear estimation could be generated by difference equations (and, later, differential equations, Kalman and Bucy, 1961). This formulation allows for nonstationary processes and a finite amount of measurement data and can be implemented in a straightforward manner with digital and analog equipment.

The advances in the area of nonlinear filtering have not been as spectacular, because the probability-density function of the state cannot be represented by a finite set of parameters (as it can in the case of Gaussian random variables). In continuous-time nonlinear estimation (problems with a nonlinear system or measurement equation) the probability-density function of the state conditioned on the available measurements must be found by solving a partial differential equation (Bucy, 1965; Kushner, 1964; Wonham, 1964). Quadratures are used to find the conditional mean and other moments.

No partial differential equation for problems in discrete-time nonlinear filtering exists, but the conditional probability-density function of the state must be updated by Bayes' rule and is usually a multidimensional integration (Ho and Lee, 1964). Some success has been obtained by linearizing the equations and then applying the theory of linear filters (Mowery, 1965). We note that this technique has little, if any, value for quantized measurements, because the slope of the nonlinearity is either zero or infinite.

With specific reference to quantized signals Bennett (1948) investigated the spectrum of the quantizer output signal when the input signal had a flat spectrum with sharp cutoff. Like Widrow (1960), he assumed the quantum intervals to be uniform and of infinite extent. Under these assumptions Widrow obtains interesting results by using the Nyquist sampling theory on probability-density functions and characteristic functions. Max (1960) investigated the optimal choice of quantizer parameters (for example, width and placement of quantum intervals), to minimize the mean square error between the quantizer input and output. Ruchkin (1961), Steiglitz (1966), and Kellog (1967) have all investigated the linear filtering of quantized signals according to various criteria. Only Kellog has considered coarse quantization.

To the best of the writer's knowledge few attempts have been made at nonlinear estimation with quantized measurements. Balakrishnan (1962) derives some results concerning an adaptive, nonlinear predictor for quantized data. Meier and his associates (1967, a), taking the Bayesian approach, use a uniform quantizer of infinite extent and derive the equations for the conditional mean and conditional covariance for a scalar state variable and only one (scalar) quantized measurement.

Estimation with quantized measurements is of prime interest to designers of digital communication systems. Three systems that have received particular attention in the past are the pulse code modulation (PCM), predictive quantization, and predictive-comparison data-compression. Studies of the quantization and reconstruction problem (PCM) have been mentioned above in connection with Ruchkin (1961), Steiglitz (1966), and Kellog (1967). Fine (1964) gives a theoretical and general treatment of optimal digital systems with an example of predictive quantization (feedback around a binary quantizer). Bello and his associates (1967) have computed Fine's nonlinear feedback function by Monte Carlo techniques and give some simulation results. Gish (1967), O'Neal (1966), and Irwin and O'Neal (1968) consider the design of linear feedback functions for the predictive-quantization system; predictive-comparison data-compression systems are in this class. Davisson (1967) treats the optimal linear feedback operation and examines an adaptive system (1966). Other approaches to data-compression have been described (*IEEE*, 1967; Davisson, 1968).

The topics on estimation with quantized measurements that are covered in this monograph are briefly summarized as follows. Chapter 2 treats the nonlinear estimation of parameter and state vectors, based on quantized measurements. The primary emphasis is placed on the determination of the minimum variance (conditional mean) estimate. Chapter 3 deals with the design of the three digital communication systems mentioned above and presents results of Monte Carlo simulation. Chapter 4 is devoted to optimal linear estimators for quantized, stationary, random processes.

1.4 Optimal Control with Quantized Measurements

The efforts of previous investigators have been in the area of linear, time-invariant, closed-loop systems that include a quantizer somewhere within the loop. Their work may be divided into two categories: deterministic and stochastic.

Typical of deterministic approaches are those taken by Bertram (1958) and by Johnson (1965), in which bounds for the system's behavior are found.

Stochastic approaches are taken by Widrow (1960), Kosyakin (1966), Graham and McRuer (1961), Smith (1966), and Gelb and Vander Velde (1968), in which the quantizer is approximated by a gain element or a noise source or both. Generally speaking, the quantum intervals are assumed to be small enough for simplifying assumptions to be made about the quantization noise as, for example, that it is white noise. System design and compensation are then carried out by using the conventional linear-design tools.

The control of systems with quantized measurements may also be pursued via the state-space techniques. A cost criterion is established, and the control actions are chosen so as to minimize the expected value of the cost. This is the problem of *optimal stochastic control*, or *combined estimation and control.* The general statement of the problem for nonlinear systems (including nonlinear measurements) and the method of solution (dynamic programming) are outlined by Fel'dbaum (1960), Dreyfus (1965, 1964) and Aoki (1967).

In only one case has the optimal stochastic control been shown to be at all practical to use or even to find. If the dynamic system and measurements are linear, the cost quadratic, the noises additive and Gaussian, and the initial conditions of the state Gaussian, then, as has been shown by Joseph and Tou (1961), the optimal stochastic control sequence does not have to be determined by dynamic programming but may be found much more easily with the aid of the separation theorem. The separation theorem states that the computation of the optimal stochastic control may be carried out in two parts: first with a Kalman filter for generating the conditional mean of the state and next with the optimal (linear) controller that is derived if all disturbances are neglected.

The topics on optimal stochastic control are as follows. Chapter 5 contains the statement of the problem and presents results for the optimal control of a two-stage process with quantized measurements; it also gives the derivation of a separation theorem for nonlinear measurements. Chapter 6 considers suboptimal stochastic control algorithms and offers a new algorithm, which is widely applicable and compares favorably with other methods in simulations; it also discusses the design of optimal linear controllers (including linear feedback for the predictive-quantization systems). Chapter 7 summarizes the results and conclusions and suggests topics for further research.

1.5 Notation

The notation follows general practice: lower case and upper case bold face letters denote vectors and matrices, respectively, e.g. $\boldsymbol{a}$ and $\boldsymbol{A}$. E is the expectation operator. Subscripts refer to the time index ($\boldsymbol{x}_n = \boldsymbol{x}(t_n)$), and superscripts refer to elements within an array ($\boldsymbol{x} = \{x^i\}$). In places where confusion might arise, the dummy arguments of probability density functions are explicitly shown with a mnemonic correlation between the dummy arguments and the random variables. For example, $p_{\boldsymbol{x},\boldsymbol{y},\boldsymbol{z}}(\boldsymbol{\xi}, \boldsymbol{\eta}, \boldsymbol{\zeta})$ is the joint probability density function of the random vectors $\boldsymbol{x}$, $\boldsymbol{y}$, and $\boldsymbol{z}$; the dummy arguments $\boldsymbol{\xi}$, $\boldsymbol{\eta}$, and $\boldsymbol{\zeta}$ refer to $\boldsymbol{x}$, $\boldsymbol{y}$, and $\boldsymbol{z}$, respectively. When the context is clear, $p(\boldsymbol{x}, \boldsymbol{y}, \boldsymbol{z})$ is used to represent the joint probability density function.

2 Nonlinear Estimation with Quantized Measurements

2.1 Introduction

This chapter contains some results in the nonlinear estimation of parameter and state vectors when the measurements are quantized. Many different criteria may be used to derive estimates of random variables (see, for example, Lee, 1964, Chapter 3), the techniques vary according to the amount of probabilistic structure assumed a priori and the amount of computation that can be performed. The next section of this chapter introduces the concept of maximum-likelihood estimation with quantized measurements. The remainder of the chapter is devoted to Bayesian estimates with emphasis on the minimum variance, or conditional mean, estimate. Estimates of parameters are considered first and lead to apparently new formulas for the conditional mean and covariance of random variables whose a priori distribution is normal. These results are extended to Gaussian, linear systems, and the Kalman filter plays a surprising role in the determination of the conditional mean. The last section describes several nonlinear approximations for calculating the conditional mean of both parameter vectors and state vectors.

2.2 Maximum-Likelihood Estimates of Parameters

In this section we derive the necessary conditions for the maximum-likelihood estimate of parameters when the observations have been quantized. This method of estimation can be used when no probabilistic structure has been imposed on the parameters. Let the measurements and the parameters be related through the linear measurement equation

$$z = \boldsymbol{H}\boldsymbol{x} + \boldsymbol{v}, \tag{2.1}$$

where $\boldsymbol{x}$ is the parameter vector, z is the measurement vector, and $\boldsymbol{v}$ is the measurement noise whose components are independent and have known distributions.

Recall that with unquantized measurements the maximum-likelihood estimate is that value of $\boldsymbol{x}$ which maximizes the likelihood function, the probability-density function of z. Let $\tilde{\boldsymbol{x}}$ be the maximum-likelihood estimate and $L(z, \boldsymbol{x})$ the likelihood function. Then

$$\tilde{\boldsymbol{x}} = \arg\left[\max_{x} L(z, \boldsymbol{x})\right] = \arg\left[\max_{x} p(z : \boldsymbol{x})\right]. \tag{2.2}$$

Here the notation $p(z : \boldsymbol{x})$ means the probability-density function of z with $\boldsymbol{x}$ as a parameter of the distribution. Observe that $\tilde{\boldsymbol{x}}$ also maximizes the probability that z lies in the hypercube defined by z and $z + dz$.

When the measurements are quantized, however, numerical values of z are not available, so that the maximization of Equation 2.2 cannot be performed. Our knowledge of the measurements is reflected in the vector inequality $\boldsymbol{a} \leq z < \boldsymbol{b}$, or

$$\{a^i \leq z^i < b^i\}, \tag{2.3}$$

where a^i and b^i are the lower and upper limits of the quantum interval in which the ith component of z is known to lie. The likelihood function to be used with quantized measurements is a generalization of the one for unquantized measurements: it is the probability that the measurements fall in the hypercube defined by Equation 2.3:

$$\begin{aligned} L(\boldsymbol{a}, \boldsymbol{b}, \boldsymbol{x}) &= P(\boldsymbol{a} \leq z < \boldsymbol{b}) = P(\boldsymbol{a} - \boldsymbol{H}\boldsymbol{x} \leq \boldsymbol{v} < \boldsymbol{b} - \boldsymbol{H}\boldsymbol{x}) \\ &= \prod_i P[a^i - (\boldsymbol{H}\boldsymbol{x})^i \leq v^i < b^i - (\boldsymbol{H}\boldsymbol{x})^i]. \end{aligned} \tag{2.4}$$

The last step follows as a result of the independence of the components of $\boldsymbol{v}$. The maximum-likelihood estimate of $\boldsymbol{x}$ with quantized measurements is thus

$$\tilde{\boldsymbol{x}} = \arg\left\{\max_{\boldsymbol{x}} \prod_i P[a^i - (\boldsymbol{H}\boldsymbol{x})^i \leq v^i < b^i - (\boldsymbol{H}\boldsymbol{x})^i]\right\}. \tag{2.5}$$

The necessary equations for a local maximum of the likelihood function are derived next. For notational convenience let

$$P_i = P[a^i - (\boldsymbol{Hx})^i \leq v^i < b^i - (\boldsymbol{Hx})^i] = \int_{a^i - (\boldsymbol{Hx})^i}^{b^i - (\boldsymbol{Hx})^i} p_{v^i}(u)\, du. \tag{2.6}$$

The necessary conditions become

$$\begin{aligned} 0 &= \frac{1}{L(\boldsymbol{a}, \boldsymbol{b}, \boldsymbol{x})}\left(\frac{\partial L(\boldsymbol{a}, \boldsymbol{b}, \boldsymbol{x})}{\partial \boldsymbol{x}}\right) = \sum_i \frac{\partial P_i/\partial \boldsymbol{x}}{P_i} \\ &= \sum_i \frac{p_{v^i}(b^i - (\boldsymbol{Hx})^i) - p_{v^i}(a^i - (\boldsymbol{Hx})^i)}{P_i} \boldsymbol{h}^i, \end{aligned} \tag{2.7}$$

where the row vector $\boldsymbol{h}^i$ is the ith row of $\boldsymbol{H}$. Under the common assumption of Gaussian observation noise these equations will be composed of exponential and error functions.

2.3 Bayesian Estimates of Parameters

The derivation of the expectation of a function of a parameter vector conditioned on quantized measurements is presented in this section. The results are quite general and will be used to formulate the estimation problem for dynamical systems by proper interpretation of the parameter and measurement vectors.

The following result is very useful for dealing with quantized random variables.

Given two random vectors $\boldsymbol{x}$ and z with joint probability-density function $p_{\boldsymbol{x},z}(\xi, \zeta)$, and given that z lies in a region denoted by A. Here the Greek letter ξ is associated with the random variable $\boldsymbol{x}$ and ζ with z. Now

$$\underset{\boldsymbol{x},z|z\in A}{p}(\xi, \zeta) = \begin{cases} 0, & \zeta \notin A, \\ \dfrac{p_{\boldsymbol{x},z}(\xi, \zeta)}{P(z \in A)}, & \zeta \in A, \end{cases} \tag{2.8}$$

where

$$P(z \in A) = \int_A d\zeta \int_{-\infty}^{\infty} d\xi\, p_{\boldsymbol{x},z}(\xi, \zeta). \tag{2.9}$$

This result may be verified by defining the event B such that

$$B = \{\boldsymbol{x}, z \mid \xi \leq \boldsymbol{x} < \xi + d\xi, \zeta \leq z < \zeta + d\zeta\} \tag{2.10}$$

and by applying Bayes' rule to the conditional event

$$P(B \mid z \in A) = \frac{P(z \in A \mid B)\, P(B)}{P(z \in A)} \tag{2.11}$$

with

$$P(z \in A \mid B) = \begin{cases} 0, & \zeta \notin A, \\ 1, & \zeta \in A. \end{cases}$$

Now we proceed to the problem of parameter estimation.

Statement of the Problem

Given:

(i) $z = \boldsymbol{h}(\boldsymbol{x}, \boldsymbol{v})$, the measurement equation,

(ii) $p_{\boldsymbol{x},\boldsymbol{v}}(\xi, \boldsymbol{v})$, the joint probability-density function of parameter and noise vectors,

(iii) $z \in A$, where A is some region (a hypercube for quantized measurements).

Find:

$E[\boldsymbol{f}(\boldsymbol{x}) \mid z \in A]$, the mean of some function of $\boldsymbol{x}$ conditioned on $z \in A$.

Result:

$$E[\boldsymbol{f}(\boldsymbol{x}) \mid z \in A] = E\{E[\boldsymbol{f}(\boldsymbol{x}) \mid z] \mid z \in A\}. \tag{2.12}$$

Proof: The joint probability-density of $\boldsymbol{x}$ and z without regard to quantized measurements is found by the usual methods (Lanning and Battin, 1956; Papoulis, 1965) from (i) and (ii) above. Their joint probability-density function conditioned on $z \in A$ is then evaluated via Equation 2.8. The two marginal probability-density functions for $\boldsymbol{x}$ and z conditioned on $z \in A$ are then evaluated:

$$\underset{\boldsymbol{x}|z\in A}{p}(\xi) = \int_A \underset{\boldsymbol{x},z|z\in A}{p}(\xi, \zeta)\, d\zeta = \frac{\int_A p_{\boldsymbol{x},z}(\xi, \zeta)\, d\zeta}{P(z \in A)}, \tag{2.13}$$

$$\underset{z|z\in A}{p}(\zeta) = \int_{-\infty}^{\infty} \underset{\boldsymbol{x},z|z\in A}{p}(\xi, \zeta)\, d\xi = \begin{cases} 0, & \zeta \notin A, \\ \dfrac{p_z(\zeta)}{P(z \in A)}, & \zeta \in A. \end{cases} \tag{2.14}$$

With these preliminaries performed, the conditional expectation of $f(x)$ becomes

$$E[f(x) \mid z \in A] = \int_{-\infty}^{\infty} d\xi f(\xi) \, p_{x|z\in A}(\xi)$$

$$= \int_{-\infty}^{\infty} d\xi f(\xi) \frac{\int_A d\zeta \, p_{x,z}(\xi, \zeta)}{P(z \in A)}, \tag{2.15}$$

where the latter result is obtained by means of Equation 2.13. Using the identity

$$p_{x,z}(\xi, \zeta) = p_{x|z}(\xi, \zeta) p_z(\zeta)$$

and assuming that the order of integration in Equation 2.15 may be reversed lead to

$$E[f(x) \mid z \in A] = \int_A d\zeta \frac{p_z(\zeta)}{P(z \in A)} \left[\int_{-\infty}^{\infty} d\xi f(\xi) \, p_{x|z}(\xi, \zeta) \right]. \tag{2.16}$$

The bracketed quantity here is the conditional mean of $f(x)$, given a measurement z, and is a function of z. The other factor is the marginal probability-density function of z conditioned on $z \in A$, Equation 2.14. Now Equation 2.16 reduces to the stated result:

$$E[f(x) \mid z \in A] = \int_A d\zeta \, p_{z|z\in A}(\zeta) E[f(x) \mid z = \zeta]$$

$$= E\{E[f(x) \mid z] \mid z \in A\}. \tag{2.17}$$

The estimation problem with quantized measurements may now be considered to consist of two operations: finding $E[f(x) \mid z]$, the conditional mean of $f(x)$ given a measurement (this function of z is the usual goal of the estimation problem without quantized measurements), and averaging this function of z conditioned on $z \in A$.

The solution to the estimation problem outlined above is most beneficial in the sense that $E[f(x) \mid z]$ has been computed for a wide variety of problems of both academic and practical interest. All that remains, then, is to average this function over $z \in A$ when the measurements are quantized.

A more general statement of this property is made available by reversing an equality in Doob (1953, p. 37). Let $\mathscr{F}_1$ and $\mathscr{F}_2$ be Borel fields such that $\mathscr{F}_1$ is a subset of $\mathscr{F}_2$. Then for any random variable y

$$E(y \mid \mathscr{F}_1) = E[E(y \mid \mathscr{F}_2) \mid \mathscr{F}_1]. \tag{2.18}$$

2.4 Gaussian Parameters

In this section we specialize the general results to estimation of Gaussian random variables. The measurement equation is assumed to be linear, and the measurement noise $\boldsymbol{v}$ to be independent of the a priori distribution of $\boldsymbol{x}$. In particular,

$$\boldsymbol{x} = N(\bar{\boldsymbol{x}}, \boldsymbol{M}),$$

$$\boldsymbol{v} = N(0, \boldsymbol{R}),$$

$$z = \boldsymbol{H}\boldsymbol{x} + \boldsymbol{v}, \qquad E(\boldsymbol{x}\boldsymbol{v}^T) = 0,$$

where $N(\ ,\ )$ denotes a normal distribution whose mean and covariance are the first and second arguments, respectively, $\boldsymbol{x}$ is an n-component vector, z and $\boldsymbol{v}$ are m-component vectors, and $\boldsymbol{H}$ is an $m \times n$ matrix.

The Conditional Mean of $\boldsymbol{x}$

Now the results of the previous section will be used to great advantage. Let $\boldsymbol{f}(\boldsymbol{x}) = \boldsymbol{x}$, and evaluate $E(\boldsymbol{x} \mid z)$, as the first step in the procedure. It is well known (Lee, 1964; Bryson and Ho, 1969; Schweppe, 1967) that

$$E(\boldsymbol{x} \mid z) = \bar{\boldsymbol{x}} + \boldsymbol{K}(z - \boldsymbol{H}\bar{\boldsymbol{x}}), \tag{2.19}$$

where $\boldsymbol{K}$ is the minimum variance gain matrix for Gaussian random variables. This matrix may be computed by either of the two following formulas (Bryson and Ho, 1969; Schweppe, 1967); the second is used when there is no observation noise:

$$\boldsymbol{K} = (\boldsymbol{M}^{-1} + \boldsymbol{H}^T\boldsymbol{R}^{-1}\boldsymbol{H})^{-1}\boldsymbol{H}^T\boldsymbol{R}^{-1}, \tag{2.20}$$

$$\boldsymbol{K} = \boldsymbol{M}\boldsymbol{H}^T(\boldsymbol{H}\boldsymbol{M}\boldsymbol{H}^T + \boldsymbol{R})^{-1}. \tag{2.21}$$

The second step in the procedure for determining the conditional mean is to average $E(\boldsymbol{x} \mid z)$ over all $z \in A$. This yields

$$E(\boldsymbol{x} \mid z \in A) = \bar{\boldsymbol{x}} + \boldsymbol{K}[E(z \mid z \in A) - \boldsymbol{H}\bar{\boldsymbol{x}}]. \tag{2.22}$$

The Conditional Covariance of $\boldsymbol{x}$

The covariance of estimation errors will be derived next, and a discussion of the results follows that.

Let $\boldsymbol{e}$ be the error in the estimate:

$$\begin{aligned} \boldsymbol{e} &= \boldsymbol{x} - E(\boldsymbol{x} \mid z \in A) \\ &= \boldsymbol{x} - E(\boldsymbol{x} \mid z) + [E(\boldsymbol{x} \mid z) - E(\boldsymbol{x} \mid z \in A)]. \end{aligned} \tag{2.23}$$

The bracketed term may be evaluated from Equations 2.19 and 2.22. Substituting the result in this equation yields

$$e = [x - E(x \mid z)] + K[z - E(z \mid z \in A)]. \tag{2.24}$$

The covariance of estimation errors $E(ee^T \mid z \in A)$ is evaluated by the two-step procedure, described in Section 2.3, with $f(x)$ equal to ee^T. The functional dependence of e on x is shown in Equation 2.24.

It is known (Kalman, 1960; Lee, 1964; Schweppe, 1967) that the distribution of x conditioned on a measurement z is $N[E(x \mid z), P]$, where

$$\begin{aligned} P &= (M^{-1} + H^T R^{-1} H)^{-1} \\ &= M - MH^T(HMH^T + R)^{-1}HM. \end{aligned} \tag{2.25}$$

This is used in the first step of the procedure, which is to calculate $E(ee^T \mid z)$:

$$E(ee^T \mid z) = P + K[z - E(z \mid z \in A)][z - E(z \mid z \in A)]^T K^T. \tag{2.26}$$

This may be verified by noting in Equation 2.24 that z is held constant and that the first term is the a posteriori estimation error for the case of linear measurements. Applying the second step in the procedure yields the desired result:

$$\operatorname{cov}(x \mid z \in A) = P + K \operatorname{cov}(z \mid z \in A) K^T. \tag{2.27}$$

Discussion

The formulas for the conditional mean and covariance, Equations 2.22 and 2.27, quite clearly reduce to the linear-measurement results as the volume of the region A goes to zero, since the conditional probability-density function of z approaches an impulse.

The expression for the conditional mean resolves a "filter paradox" for quantized measurements. Filters operate on numbers, the numerical realizations of the measurements, yet a quantizer is really telling us that the measurements lie within certain bounds; that is, we have intervals, not numbers, for measurement quantities. Equation 2.22 shows that we may find the conditional mean with a linear filter and, moreover, the Gaussian linear filter, or Kalman filter, if we agree to use the conditional mean of the measurement vector as the numerical input. The onus of designing filters has been transferred to determining the proper input sequence, and this concept will be adopted frequently herein.

The conditional covariance of the estimate, Equation 2.27, depends on the particular realization of the measurements, because the covariance of z is conditioned on $z \in A$. This supports the intuitive notion that measurements lying in "small" quantum intervals provide more accurate estimates than measurements lying in "large" quantum intervals.

Another interesting result is that quantization increases the covariance of the estimate as though it were process noise added *after* the incorporation of a linear measurement. This is in sharp distinction to the usual interpretation, in which the effects of quantization are treated as observation noise added *before* a linear measurement.

The discussion of the results obtained in this section is continued in the next in the light of estimating the output of a linear dynamical system driven by Gaussian noise.

2.5 Extension to Linear Dynamic Systems: Filtering, Predicting, and Smoothing

In this section we investigate the problem of filtering, predicting, and smoothing the output of a linear dynamical system. Our objective is to derive the mean of the state vector at some time conditioned on quantized measurements. It is desirable, furthermore, to have an estimator that can accommodate nonstationary problems, i.e. time-varying systems, time-varying noise statistics, and a finite amount of data.

The Linear System

It is assumed that the state vector and measurement variables obey the relationships

$$\boldsymbol{x}_{i+1} = \boldsymbol{\Phi}_i \boldsymbol{x}_i + \boldsymbol{w}_i, \tag{2.28}$$

$$z_i = \boldsymbol{H}_i \boldsymbol{x}_i + \boldsymbol{v}_i, \tag{2.29}$$

$$E(\boldsymbol{x}_0) = \bar{\boldsymbol{x}}_0, \quad \operatorname{cov}(\boldsymbol{x}_0) = \boldsymbol{P}_0, \tag{2.30}$$

$$E(\boldsymbol{w}_i) = 0, \quad E(\boldsymbol{w}_i \boldsymbol{w}_j^T) = \boldsymbol{Q}_i \delta_{ij}, \tag{2.31}$$

$$E(\boldsymbol{v}_i) = 0, \quad E(\boldsymbol{v}_i \boldsymbol{v}_j^T) = \boldsymbol{R}_i \delta_{ij}, \tag{2.32}$$

$$E(\boldsymbol{w}_i \boldsymbol{v}_j^T) = 0, \quad E(\boldsymbol{w}_i \boldsymbol{x}_0^T) = E(\boldsymbol{v}_i \boldsymbol{x}_0^T) = 0, \tag{2.33}$$

where

$\boldsymbol{x}_i$ = system state vector at time t_i,
$\boldsymbol{\Phi}_i$ = system transition matrix from time t_i to t_{i+1}
$\boldsymbol{w}_i$ = realization of the process noise at t_i,
$\boldsymbol{z}_i$ = measurement vector at time t_i,
$\boldsymbol{H}_i$ = measurement matrix at time t_i,
$\boldsymbol{v}_i$ = realization of the observation noise at time t_i.

It is further assumed that the noises and initial conditions are normally distributed, a fact which, with the linear equations, ensures that all the random variables will be normal.

Solution to the Filtering, Predicting, and Smoothing Problems

The filtering problem is that of determining the mean of a state vector conditioned on the present and past measurements, the prediction problem is that of determining one conditioned only on past measurements, and the smoothing problem is that of determining one conditioned on present, past, and future measurements (a "future" measurement occurs at a time later than that of the state vector in question).

The solution to these three problems is given immediately by means of the conditional mean of a parameter vector, derived in the last section:

1. Let the state vector(s) under consideration be the parameter vector $\boldsymbol{x}$. Collect all the quantized measurements into the vector $\boldsymbol{z}$. Let A be the region in which these quantized measurements fall.
2. Compute the mean and covariance of $\boldsymbol{z}$ conditioned on $\boldsymbol{z} \in A$.
3. Solve Equation 2.22 for the mean (and Equation 2.27 for the covariance) of the state vector(s) conditioned on $\boldsymbol{z} \in A$. These yield the desired results, since $\boldsymbol{x}$ and $\boldsymbol{z}$ are Gaussian random variables.

Remarks

1. Equation 2.22 represents the "batch-processing" solution to the estimation problem (Lee, 1964; Schweppe, 1967). Once the mean of $\boldsymbol{z}$, given $\boldsymbol{z} \in A$, is in hand, however, the equations may be solved recursively via the Kalman filter (Lee, 1964; Schweppe, 1967); it may be noted that with Equation 2.18 the conditional mean for continuous-time systems can be generated with the continuous-time Kalman filter in the same way. The input to the filter at time t_i is $E(\boldsymbol{z}_i \mid \boldsymbol{z} \in A)$.

2. This formulation, like the Kalman filter, produces the conditional mean, even with time-varying systems, time-varying noise statistics, and a finite number of measurements.

3. It has not been assumed that the quantum intervals are "small." Nor has any restriction been placed on the quantizer; that is, it may be nonuniform and may have any number of output levels from two to infinity. Most analyses rely on one or more assumptions like these.

4. The measurements at different times are correlated; consequently, the conditional mean of, say, z_i will change as more measurements are taken. Thus, even though the equations may be solved recursively, $E(z \mid z \in A)$ must be recomputed after each measurement; this necessitates a complete solution of Equation 2.22 after each measurement is taken.

5. The dimension of z is enlarged with each new measurement, so that the memory requirements are ever increasing. This could be remedied by retaining only a finite amount of data for estimation.

6. The conditional covariance, Equation 2.27, depends on the measurement sequence and cannot be computed in advance, as it can with the Kalman filter. For a stationary process, then, the conditional covariance should not be expected to reach a steady-state value, even though in the Gaussian linear case it may approach a constant.

In sum, the conditional mean (and covariance) of a nonstationary Gaussian linear system has been found when the measurements are quantized. The solution resolves the filter paradox of "what numbers should be used as input and how they should be weighted": the input sequence is the conditional mean of the measurement vector, and this is used to drive the Kalman filter that would have been used had the measurements been linear.

The primary advantages of the present approach are that the accuracy in the computation of the conditional mean of x is limited only by the accuracy in determining the conditional mean of $z \in A$ and that no assumptions about the number of quantum levels or their widths or locations have been made. This allows one to consider very general quantization schemes. Unlike the Kalman filter, however, the entire input sequence must be redetermined after each new measurement. This in turn requires a rerunning of the Kalman filter from its specified initial conditions. An unbounded amount of storage is required, although the formulation given above may be used for a truncated memory. The most time-consuming aspect of the computation is the determination of the conditional mean and covariance of the measurement vector; it is to this problem that we now turn.

2.6 Conditional Mean and Covariance of the Measurement Vector

In this section we consider some methods of evaluating the mean and covariance of the measurement vector conditioned on the fact that the components have been quantized. The mean and covariance thus obtained may be used in conjunction with the Kalman filter to determine the conditional mean and covariance of the desired state vector(s) for Gaussian linear systems.

Integral Expressions

For a computation of the conditional mean and covariance the region A in which the measurement vector lies must be defined more precisely. Let the vector inequality

$$\boldsymbol{a} \leq z < \boldsymbol{b} \tag{2.34}$$

imply

$$a^i \leq z^i < b^i, \qquad i = 1, 2, \ldots, m. \tag{2.35}$$

The conditional probability-density function of the measurement vector has already been found in Equation 2.14: it has the same functional form as the original probability-density function over the region A. The conditional mean may thus be expressed by the following multidimensional integrals:

$$E(z \mid z \in A) = \frac{\int_{\boldsymbol{a}}^{\boldsymbol{b}} \zeta\, p_z(\zeta)\, d\zeta}{\int_{\boldsymbol{a}}^{\boldsymbol{b}} p_z(\zeta)\, d\zeta}. \tag{2.36}$$

A similar formula may be written for the conditional covariance:

$$\operatorname{cov}(z \mid z \in A) = \frac{\int_{\boldsymbol{a}}^{\boldsymbol{b}} [\zeta - E(z \mid z \in A)][\zeta - E(z \mid z \in A)]^T p_z(\zeta)\, d\zeta}{\int_{\boldsymbol{a}}^{\boldsymbol{b}} p_z(\zeta)\, d\zeta}. \tag{2.37}$$

Note that the semi-infinite end intervals associated with every real quantizer may easily be treated, since the quantities a^i and b^i may take the values $-\infty$ and $+\infty$, respectively.

Another interpretation of the filter operation (computing an estimate as a function of numerical data) is as follows. Since Equations 2.36 and

2.37 are functions of the numbers $\boldsymbol{a}$ and $\boldsymbol{b}$, the conditional mean and covariance are functionally related to these numbers, too. Thus, each quantized scalar measurement requires *two* numbers for the filter input: these numbers are the bounds of the quantum interval in which the measurement lies (disjoint quantum intervals may also be treated in this formulation, but they require two more filter inputs for each additional disjoint interval).

If there are m components in the measurement vector, the number of individual m-dimensional integrations required for computing the conditional mean and covariance is

$$1 + m + \frac{m(m + 1)}{2}.$$

The first term is the scaling factor $P(z \in A)$, the second is the number of components in the conditional mean, and the third is the number required for the (symmetric) conditional covariance matrix. No known analytical form exists for the solutions to the conditional mean and covariance, Equations 2.36 and 2.37, for multivariate normal distributions. Quite a bit of effort has gone into just finding approximations to $P(z \in A)$; see Gupta (1963). Analytical progress is limited to special cases of correlation coefficients, and no mention is made of the conditional mean or covariance.

The Scalar Gaussian Case

In the scalar case the following relationships hold if z is Gaussian with mean $\bar{z}$ and variance σ_z^2:

$$E(z \mid z \in A) = \bar{z} + \frac{\sigma_z}{P(z \in A)} \times \left(\frac{\exp\{-\frac{1}{2}[(a - \bar{z})/\sigma_z]^2\}}{(2\pi)^{1/2}} - \frac{\exp\{-\frac{1}{2}[(b - \bar{z})/\sigma_z]^2\}}{(2\pi)^{1/2}}\right), \tag{2.38}$$

$$\operatorname{cov}(z \mid z \in A) = \sigma_z^2\left[1 + \left(\frac{a - \bar{z}}{\sigma_z}\right)\frac{\exp\{-\frac{1}{2}[(a - \bar{z})/\sigma_z]^2\}}{(2\pi)^{1/2}P(z \in A)} - \left(\frac{b - \bar{z}}{\sigma_z}\right)\frac{\exp\{-\frac{1}{2}[(b - \bar{z})/\sigma_z]^2\}}{(2\pi)^{1/2}P(z \in A)} - \frac{1}{P(z \in A)^2} \times \left(\frac{\exp\{-\frac{1}{2}[(a - \bar{z})/\sigma_z]^2\}}{(2\pi)^{1/2}} - \frac{\exp\{-\frac{1}{2}[(b - \bar{z})/\sigma_z]^2\}}{(2\pi)^{1/2}}\right)^2\right], \tag{2.39}$$

where

$$P(z \in A) = \int_{(a-\bar{z})/\sigma_z}^{(b-\bar{z})/\sigma_z} \frac{\exp(-\frac{1}{2}v^2)}{(2\pi)^{1/2}} \, dv \tag{2.40}$$

$$= \frac{1}{2}\left[\operatorname{erf}\left(\frac{b-\bar{z}}{2^{1/2}\sigma_z}\right) - \operatorname{erf}\left(\frac{a-\bar{z}}{2^{1/2}\sigma_z}\right)\right]. \tag{2.41}$$

Independent Measurements

If the components of the Gaussian measurement vector are uncorrelated and thus independent of each other, the computation of the conditional mean and covariance of the measurement vector is greatly reduced. In this case the marginal probability-density functions of the individual components conditioned on $\mathbf{z} \in A$ are not influenced by the other measurements, and the formulas for the scalar random variables, Equations 2.38 and 2.39, may be applied to each individual component. Unfortunately, for dynamic systems the state vectors at different time instants are correlated, and this in turn implies that the measurements at those times also are correlated.

A reasonable question to ask at this point is the following: using the a priori information available, why not "whiten" the measurement vector—that is, make the linear transformation to independent Gaussian random variables—and *then* pass the measurements through the quantizer? The answer is that such a procedure would defeat the purpose of processing the data, since it is the correlation between samples that allows one to get a better idea of just where the sample lies within the quantum interval.

Block Quantization

Another scheme for quantizing correlated measurements has been proposed by Huang and Schultheiss (1962). A block of, say, m Gaussian measurements is collected and then orthogonally transformed into m independent variables. Each of these new random variables is quantized *in a different quantizer* under the constraint that the total number of bits (or output levels) for the block is fixed. The greatest number of levels is assigned to the independent random variable with the largest variance. The advantages of this method are the following:

1. The "whitening" filter can be determined a priori.
2. The conditional mean and covariance are easily computed from Equations 2.38 and 2.39.
3. The ensemble covariance has already been tabulated.

4. The method is relatively easy to use in analytic manipulations.

The disadvantages are primarily operational:

1. Real-time operation cannot be carried out, because all data in the block must be taken before quantization is performed.
2. Additional computation is required to whiten the data.
3. If the data are coming from different sources, and any one source produces spurious data (as has been known to happen), it will have a detrimental effect on the determination of other measurements.

There is one situation in which the block quantization scheme would be useful: when vector measurements of a linear system are taken at each observation time. In this case all the data arrive at the same time, and the data may be transformed and quantized without delay.

Least Squares

If the quantizer is assigned a fixed input-output relationship, then a least-squares linear fit of the quantizer outputs may be made, to yield an approximation to the conditional mean of the measurement vector. This results in a linear operation that is not the optimal linear filter.

The method gives very poor results, because it neglects the inter-sample correlation of the quantization "noise" (quantization noise is the difference between the quantizer output and input). The optimal linear filter accounts for this correlation; it is treated in detail in Chapter 4.

Numerical Integration

Equation 2.36 for the conditional mean can be integrated numerically by using quadrature formulas in each of the integration variables (Klerer and Korn, 1967). The primary disadvantage here is the amount of computer time required, because the integrand must be evaluated many times: M points in each of m dimensions require M^m evaluations of the integrand. For an almost minimal number of points, say $M = 4$, and a memory length of, say, $m = 10$ it would require approximately 40 seconds on an IBM 360 Model 65 just to evaluate the exponential function $4^{10} \approx 10^6$ times. This technique could have applications when the number of dimensions is small, on the order of 2 or 3; for $M = 8$ points in each of three dimensions, or $m = 3$, the number of evaluations is reduced to $8^3 = 512$, which requires only 0.02 second of computation. Algorithms that compute the more important measurements with greater accuracy will save computer time without appreciable loss of accuracy.

Integration formulas that require fewer evaluations of the integrand are available (Klerer and Korn, 1967). They are based on approximating the integrand by a linear combination of product polynomials of the form

$$z_1^{\alpha_1} z_2^{\alpha_2} \cdots z_m^{\alpha_m}, \qquad \alpha_1 + \alpha_2 + \cdots + \alpha_m \leq k, \quad \alpha_i \geq 0, \tag{2.42}$$

where k is the degree of precision of the integration routine. The range of integration is normalized to a unit hypercube. The coefficients of the terms in this equation are found by evaluating the integrand at points such that, if it were a polynomial of degree k or less, the integration formulas would be exact. These formulas were tried with memory lengths of 5 to 10 and were found to be both time-consuming and inaccurate.

2.7 Approximate Nonlinear Estimation

Power-Series Expansion

This section contains a power-series method of approximating the conditional mean and covariance of the measurement vector. The method is limited in use to cases in which the ratio of quantum interval to standard deviation is small, but it is very flexible and can "adapt" to changing quantum intervals.

Each of the m components of the normally distributed vector $\boldsymbol{z}$ has zero mean and lies in an interval whose limits are $\{a^i\}$ and $\{b^i\}$,

$$a^i \leq z^i < b^i, \qquad i = 1, 2, \ldots, m \tag{2.43}$$

or $\boldsymbol{a} \leq \boldsymbol{z} < \boldsymbol{b}$. The geometric center of the region A defined by Equation 2.43 is the vector $\boldsymbol{\gamma}$,

$$\boldsymbol{\gamma} = \tfrac{1}{2}(\boldsymbol{b} + \boldsymbol{a}) \tag{2.44}$$

and the vector of quantum interval halfwidths is $\{\alpha^i\}$,

$$\boldsymbol{\alpha} = \tfrac{1}{2}(\boldsymbol{b} - \boldsymbol{a}). \tag{2.45}$$

The probability-density function is expanded in a power series about $\boldsymbol{\gamma}$, and terms of fourth order and higher are neglected. The details are carried out in Appendix A; the result is that the mean and covariance of $\boldsymbol{z}$ conditioned on $\boldsymbol{z} \in A$ are given by

$$E(\boldsymbol{z} \mid \boldsymbol{z} \in A) \approx \boldsymbol{\gamma} - \boldsymbol{A}\boldsymbol{\Gamma}^{-1}\boldsymbol{\gamma} \tag{2.46}$$

$$\operatorname{cov}(\boldsymbol{z} \mid \boldsymbol{z} \in A) \approx \boldsymbol{A} = \left\{ \frac{(\alpha^i)^2}{3} \delta_{ij} \right\} \tag{2.47}$$

where Γ is $E(zz^T)$ and δ_{ij} is the Kronecker delta. The conditional mean is the center of the region plus a second-order correction term. The weighting matrix of this second term depends on the individual quantum-interval widths given by Equation 2.47. The conditional covariance is the same as it would be if the probability density were uniformly distributed even though terms as high as third-order were retained in the expansion.

Because this case involves "small" quantum intervals, it is informative to compare this approximation to the conditional mean with the commonly used method, which assumes that the quantizer is an additive-noise source (Widrow, 1960; Ruchkin, 1961; Steiglitz, 1966). The quantizer noise, defined as the quantizer output minus its input, has a covariance matrix $\boldsymbol{A}$ given by Equation 2.47. *After* the observations are made and the α^i are known, the minimum-variance linear estimate $\boldsymbol{x}^*$ and its covariance $\boldsymbol{E}^*$ are given by

$$\boldsymbol{x}^* = \bar{\boldsymbol{x}} + \boldsymbol{K}^*(\gamma - \boldsymbol{H}\bar{\boldsymbol{x}}), \tag{2.48}$$

$$\boldsymbol{E}^* = \boldsymbol{M} - \boldsymbol{M}\boldsymbol{H}^T(\boldsymbol{\Gamma} + \boldsymbol{A})^{-1}\boldsymbol{H}\boldsymbol{M}, \tag{2.49}$$

where

$$\boldsymbol{K}^* = \boldsymbol{M}\boldsymbol{H}^T(\boldsymbol{\Gamma} + \boldsymbol{A})^{-1}, \tag{2.50}$$

$$\boldsymbol{\Gamma} = \operatorname{cov}(z) = \boldsymbol{H}\boldsymbol{M}\boldsymbol{H}^T + \boldsymbol{R}. \tag{2.51}$$

These equations can be solved recursively with the Kalman filter when they are equivalent to the batch-processing estimate for a dynamic system described by Equations 2.28 to 2.33. It is still a nonlinear estimate, however, since the α^i are functions of the observations; filter weights cannot be computed in advance unless the α^i are all equal.

To compare the estimate $\boldsymbol{x}^*$ based on the quantizer-noise model with the approximate conditional mean, rewrite Equation 2.48 as

$$\boldsymbol{x}^* = \bar{\boldsymbol{x}} + \boldsymbol{M}\boldsymbol{H}^T\boldsymbol{\Gamma}^{-1}(\boldsymbol{I} + \boldsymbol{A}\boldsymbol{\Gamma}^{-1})^{-1}(\gamma - \boldsymbol{H}\bar{\boldsymbol{x}}). \tag{2.52}$$

Expand the matrix $(\boldsymbol{I} + \boldsymbol{A}\boldsymbol{\Gamma}^{-1})^{-1}$ in a power series, and neglect fourth and higher order powers of the ratio of quantum interval to standard deviation or, equivalently, $(\boldsymbol{A}\boldsymbol{\Gamma}^{-1})^n$, with $n \geq 2$. [A necessary and sufficient condition that this series converge is that the largest eigenvalue of $\boldsymbol{A}\boldsymbol{\Gamma}^{-1}$ be less than 1 (Wilkinson, 1965, p. 59); a sufficient condition, with the use of matrix norms, is that the largest value of $(\alpha^i)^2/3$ is less than the smallest eigenvalue of $\boldsymbol{\Gamma}$.] After terms of higher order are dropped, Equation 2.52 becomes

$$\boldsymbol{x}^* \approx \bar{\boldsymbol{x}} + \boldsymbol{K}(\boldsymbol{I} - \boldsymbol{A}\boldsymbol{\Gamma}^{-1})(\gamma - \boldsymbol{H}\bar{\boldsymbol{x}}), \tag{2.53}$$

where $\boldsymbol{K} = \boldsymbol{M}\boldsymbol{H}^T\boldsymbol{\Gamma}^{-1}$ is the optimal gain for linear measurements. An approximate expression for the conditional mean is found by substituting Equation 2.46 in Equation 2.22 and accounting for a nonzero mean of z. This yields

$$E(\boldsymbol{x} \mid z \in A) \approx \bar{\boldsymbol{x}} + \boldsymbol{K}(\boldsymbol{I} - A\boldsymbol{\Gamma}^{-1})(\gamma - \boldsymbol{H}\bar{\boldsymbol{x}}). \tag{2.54}$$

Thus the two estimates agree within the approximations that have been made. Similar calculations show that

$$\operatorname{cov}(\boldsymbol{x} \mid z \in A) = \boldsymbol{E}^* + \text{terms of fourth and higher order.} \tag{2.55}$$

This equation is quantitative verification of a well-known fact: good results can be obtained when the quantizer is considered a noise source, if the quantum intervals are small. Moreover, it can be used to predict just how well the noise-model filter will perform relative to the conditional-mean estimate. For example, if the ratio of threshold halfwidth to standard deviation is 0.3 (one quantum interval is 0.6σ), then $\boldsymbol{E}^*$ should be about $0.3^4 = 0.01$, or 1 percent larger than $\operatorname{cov}(\boldsymbol{x} \mid z \in A)$.

The Gaussian Fit Algorithm

The Gaussian fit algorithm is the writer's name for a discrete-time nonlinear filter that recursively fits a Gaussian distribution to the first two moments of the *conditional* distribution of a system state vector. It is analogous to the suggestion of Jazwinski (1966), who treated discrete measurements of a continuous-time nonlinear state equation. Bass and Schwartz (1966) examine an equation of state containing both continuous-time measurements and dynamics; they expand the nonlinear measurement function in a power series—a procedure that is inapplicable to quantization. Fisher (1966) apparently matches more generalized moments of the conditional distribution. Davisson (1967) makes similar assumptions concerning the distribution over the *ensemble* of measurements. Here we present a heuristic justification for the technique, and we derive (in Appendix B) an ensemble average performance estimate for stationary or nonstationary data.

Consider the system described by Equations 2.28 to 2.33, and assume the following.

Assumption: The conditional distribution of the state just prior to the ith measurement is $N(\hat{\boldsymbol{x}}_{i|i-1}, \boldsymbol{M}_i)$.

Then we know from Section 2.4 and Equations 2.28 to 2.33 that

$$\hat{\boldsymbol{x}}_{i|i} = \hat{\boldsymbol{x}}_{i|i-1} + \boldsymbol{K}_i[E(z_i \mid z_i \in A_i) - \boldsymbol{H}_i\hat{\boldsymbol{x}}_{i|i-1}], \tag{2.56}$$

$$\boldsymbol{K}_i = \boldsymbol{M}_i\boldsymbol{H}_i^T(\boldsymbol{H}_i\boldsymbol{M}_i\boldsymbol{H}_i^T + \boldsymbol{R}_i)^{-1}, \tag{2.57}$$

$$\boldsymbol{P}_i = \boldsymbol{M}_i - \boldsymbol{M}_i\boldsymbol{H}_i^T(\boldsymbol{H}_i\boldsymbol{M}_i\boldsymbol{H}_i^T + \boldsymbol{R}_i)^{-1}\boldsymbol{H}_i\boldsymbol{M}_i, \tag{2.58}$$

$$\boldsymbol{E}_i = \boldsymbol{P}_i + \boldsymbol{K}_i \operatorname{cov}(z_i \mid z_i \in A_i)\boldsymbol{K}_i^T, \tag{2.59}$$

$$\hat{\boldsymbol{x}}_{i+1|i} = \boldsymbol{\Phi}_i\hat{\boldsymbol{x}}_{i|i}, \tag{2.60}$$

$$\boldsymbol{M}_{i+1} = \boldsymbol{\Phi}_i\boldsymbol{E}_i\boldsymbol{\Phi}_i^T + \boldsymbol{Q}_i, \tag{2.61}$$

where

$\hat{\boldsymbol{x}}_{i|i}$ = conditional mean (under the assumption) of $\boldsymbol{x}_i$ given quantized measurements up to and including t_i,
$\hat{\boldsymbol{x}}_{i|i-1}$ = conditional mean (under the assumption) of $\boldsymbol{x}_i$ given quantized measurements up to and including t_{i-1},
A_i = quantum region in which z_i falls,
$\boldsymbol{M}_i$ = conditional covariance (under the assumption) of $\boldsymbol{x}_i$, given quantized measurements up to and including t_{i-1},
$\boldsymbol{K}_i$ = Kalman-filter gain matrix at t_i,
$\boldsymbol{P}_i$ = conditional covariance (under the assumption) of estimate, had the ith measurement been linear,
$\boldsymbol{E}_i$ = conditional covariance (under the assumption) of $\boldsymbol{x}_i$, given quantized measurements up to and including t_i.

Under the assumption given above, Equations 2.60 and 2.61 correctly describe the propagation of the first two moments of the conditional distribution, although it is no longer Gaussian. The Gaussian fit algorithm assumes that the assumption is again true at time t_{i+1}; in other words, it "fits" a Gaussian distribution to the moments given by these two equations. To give some justification for this procedure, let $\boldsymbol{e} = \boldsymbol{x} - \hat{\boldsymbol{x}}$, and subtract Equation 2.60 from the state equation, Equation 2.28:

$$\boldsymbol{e}_{i+1|i} = \boldsymbol{\Phi}_i\boldsymbol{e}_{i|i} + \boldsymbol{w}_i. \tag{2.62}$$

Since $\boldsymbol{e}_{i|i}$ is not Gaussian, $\boldsymbol{e}_{i+1|i}$ is not Gaussian, either, although it should tend toward a Gaussian distribution in the majority of cases because of the addition of Gaussian process noise $\boldsymbol{w}_i$ and the mixing of the components of $\boldsymbol{e}_{i|i}$ by the state transition matrix.

Because the assumption given above is not exact, the Gaussian fit algorithm described by the recursion relations Equations 2.56 to 2.61 yields only approximations to the conditional moments. These recursion relations are very much like the Kalman filter, with two important differences:

1. The conditional mean of the measurement vector at t_i is used as the filter input; this conditional mean is computed on the assumption that the distribution of the measurement is $N(\boldsymbol{H}_i\hat{\boldsymbol{x}}_{i|i-1}, \boldsymbol{H}_i\boldsymbol{M}_i\boldsymbol{H}_i^T + \boldsymbol{R}_i)$.

2. The conditional covariance, Equation 2.59, is being forced by the random variable $\text{cov}(z_i \mid z_i \in A_i)$; in general there is no steady-state mean square error for stationary input processes, and the filter weights are random until the previous measurement has been taken.

The primary advantages of the Gaussian fit algorithm are that it is relatively easy to compute, that it can handle nonstationary data as easily as stationary data, and that its general operation is independent of the quantization scheme used. The primary disadvantages are that it requires more computation than the optimal linear filter and that it can be applied with some justification only to Gauss–Markov processes.

Performance Estimates The difficulty in analyzing the ensemble average performance of the Gaussian fit algorithm is due to the fact that filter weights are random variables, since they are functions of past measurements. Although one result of the filter computations is an approximation to the conditional covariance, this is not the ensemble covariance, which is obtained by averaging over all possible measurement sequences (for linear measurements, however, these two covariances are the same). Approximate performance estimates are derived in Appendix B for the three systems considered in the next chapter.

Gaussian Fit Smoothing

In this section we consider the problem of obtaining smoothed estimates of the state vector of a linear system driven by Gaussian noise. It is assumed that measurements have been taken from time t_1 to t_K and now it is desired to obtain estimates of the state vectors at times t_0 to t_K on the basis of *all* these data.

The technique to be used here is analogous to that of Fraser (1967). He shows that the optimal smoothed estimate based on linear measurements can be found by combining the outputs from two linear filters: the usual Kalman filter running forward from the specified initial conditions of mean and covariance and an optimal linear filter running backward from the terminal time; the latter filter starts with no a priori information.

Slight modifications to Fraser's method must be made, to take full advantage of the Gaussian fit hypothesis, but the procedure may be briefly described as follows. The Gaussian fit filter is run from time t_0 to generate a prediction of $\boldsymbol{x}_k$ based on quantized measurements at $t_1, t_2, \ldots, t_{k-1}$. Another Gaussian fit filter is run backward from time t_K, and another prediction of $\boldsymbol{x}_k$ is made on the basis of quantized measurements at times $t_{k+1}, \ldots, t_K$. The errors in both these estimates are Gaussian by hypothesis, and the two estimates are combined to give a Gaussian estimate of the state at t_k that is based on all measurements except the one at t_k. The measurement at t_k is then incorporated by using the formulas for the conditional mean and covariance of Gaussian parameters derived in Section 2.4. The measurement at t_k could, in theory, be accounted for in one of the filters before the estimates are combined. This combination would be much more difficult, because the estimate that includes the measurement at t_k is no longer Gaussian.

Let $\hat{\boldsymbol{x}}_{k|k-1}$ and $\boldsymbol{M}_k$ be the mean and covariance of the state at time t_k conditioned on quantized measurements up to and including t_{k-1}. The error in this estimate is assumed to be Gaussian. Let $\hat{\boldsymbol{x}}_{k|K-k}$ and $\boldsymbol{M}_{K-k}$ be the mean and covariance of the state at time t_k, conditioned on measurements $t_{k+1}, \ldots, t_K$ and no a priori information. This estimate, which also is computed via the Gaussian fit algorithm, is presumed to have a normally distributed error, too. It can be shown that the error in each of these estimates is a function of the initial estimation error in each of the estimates and the realizations of the process-noise and observation-noise vectors. Since the two initial estimation errors and the noise vectors are assumed independent of each other, it follows that the errors in the two estimates $\hat{\boldsymbol{x}}_{k|k-1}$ and $\hat{\boldsymbol{x}}_{k|K-k}$ are independent of each other. These two estimates are combined to give the mean of $\boldsymbol{x}_k$ conditioned on quantized measurements at $t_1, \ldots, t_{k-1}, t_{k+1}, \ldots, t_K$; that is, all measurements have been accounted for except that at time t_k. Let $\hat{\boldsymbol{x}}'_{k|K}$ be the conditional mean of this combined estimate and $\boldsymbol{M}_{k|K}$ its conditional covariance. These quantities (Fraser, 1967) are

$$\hat{\boldsymbol{x}}'_{k|k} = \boldsymbol{M}_{k|K}(\boldsymbol{M}_k^{-1}\hat{\boldsymbol{x}}_{k|k-1} + \boldsymbol{M}_{K-k}^{-1}\hat{\boldsymbol{x}}_{k|K-k}), \tag{2.63}$$

$$\boldsymbol{M}_{k|K}^{-1} = \boldsymbol{M}_k^{-1} + \boldsymbol{M}_{K-k}^{-1}. \tag{2.64}$$

Now the quantized measurement at time t_k is incorporated by using the formulas for the conditional mean and covariance of Gaussian parameters derived in Section 2.4. The a priori mean and covariance that are to be used in these expressions are given by Equations 2.63 and 2.64, respectively. Let $\hat{x}_{k|K}$ and $E_{k|K}$ be the smoothed estimate and its covariance. From Equations 2.22 and 2.27 these quantities are found to be

$$\hat{x}_{k|K} = \hat{x}'_{k|K} + K_{k|K}[E(z_k \mid \hat{x}'_{k|K}, M_{k|K}, A_k) - H_k\hat{x}'_{k|K}], \tag{2.65}$$

$$E_{k|K} = P_{k|K} + K_{k|K} \operatorname{cov}(z_k \mid \hat{x}'_{k|K}, M_{k|K}, A_k)K^T_{k|K}, \tag{2.66}$$

where A_k is the region in which z_k lies, and

$$P_{k|K} = M_{k|K} - M_{k|K}H_k^T(H_kM_{k|K}H_k^T + R_k)^{-1}H_kM_{k|K}, \tag{2.67}$$

$$K_{k|K} = M_{k|K}H_k^T(H_kM_{k|K}H_k^T + R_k)^{-1}. \tag{2.68}$$

The fact that the mean and covariance of the measurement vector in Equations 2.65 and 2.66 are conditioned on $\hat{x}'_{k|K}$ and $M_{k|K}$ means that these two quantities define the Gaussian distribution from which the conditional mean and covariance are to be determined.

Starting the Gaussian Fit Filter with No A Priori Information The equations for the backward Gaussian fit filter can be found by using the state difference equation solved backward in time (see Fraser, 1967, for filters which use other variables). The operation of this filter is identical with that of the forward filter when the conditional-covariance matrix exists. However, the backward filter starts out with infinite covariance (zero information). In this section are described several alterations of the backward filter, to account for the case in which the conditional covariance does not exist. Temporarily assume that the inverses of all matrices exist, and define quantities for the backward Gaussian fit filter as follows:

M_{K-k} = covariance of prediction of x_k based on measurements at $t_{k+1}, \ldots, t_K$,

E_{K-k} = covariance of estimate of x_k based on measurements at $t_k, t_{k+1}, \ldots, t_K$,

$$F_{K-k} = H_kM_{K-k}H_k^T, \tag{2.69}$$

$$\Gamma_{K-k} = H_kM_{K-k}H_k^T + R_k = F_{K-k} + R_k, \tag{2.70}$$

$$K_{K-k} = M_{K-k}H_k^T\Gamma_{K-k}^{-1}, \tag{2.71}$$

$$P_{K-k} = M_{K-k} - M_{K-k}H_k^T\Gamma_{K-k}^{-1}H_kM_{K-k}. \tag{2.72}$$

THE CONDITIONAL MEAN OF z. The conditional mean of the measurement for a small ratio of quantum interval to standard deviation is given by Equation A.14 in Appendix A,

$$E(z_k \mid z_k \in A_k) \approx \gamma_k - A_k \mathbf{\Gamma}_{K-k}^{-1} \gamma_k, \tag{2.73}$$

where γ_k is the geometric center of the quantum region A_k. If the norm of $\mathbf{\Gamma}_{K-k}^{-1}$ approaches zero when $\boldsymbol{M}_{K-k}$ ceases to exist, then this equation becomes

$$\lim_{\mathbf{\Gamma}_{K-k}^{-1} \to 0} E(z_k \mid z_k \in A_k) = \gamma_k.$$

PROPAGATION OF THE INFORMATION MATRIX. If the covariances in the backward filter exist, they obey the following equation between measurements:

$$\boldsymbol{M}_{K-k+1} = \boldsymbol{\Phi}_k^{-1}(\boldsymbol{E}_{K-k} + \boldsymbol{Q}_k)(\boldsymbol{\Phi}_k^{-1})^T, \qquad k = K, K-1, \ldots, 1. \tag{2.74}$$

Now invert this equation, and apply the well-known matrix identity

$$(\boldsymbol{M}^{-1} + \boldsymbol{H}^T \boldsymbol{R}^{-1} \boldsymbol{H})^{-1} = \boldsymbol{M} - \boldsymbol{M}\boldsymbol{H}^T(\boldsymbol{H}\boldsymbol{M}\boldsymbol{H}^T + \boldsymbol{R})^{-1}\boldsymbol{H}\boldsymbol{M} \tag{2.75}$$

to the right-hand side of Equation 2.74:

$$\boldsymbol{M}_{K-k+1}^{-1} = \boldsymbol{\Phi}_k^T[\boldsymbol{Q}_k^{-1} - \boldsymbol{Q}_k^{-1}(\boldsymbol{Q}_k^{-1} + \boldsymbol{E}_{K-k}^{-1})^{-1}\boldsymbol{Q}_k^{-1}]\boldsymbol{\Phi}_k. \tag{2.76}$$

This equation describes the propagation of the information matrix between measurements.

UPDATING THE INFORMATION MATRIX. The information matrix, after the quantized measurement has been processed, is found by inverting the covariance matrix of estimation error, $\boldsymbol{E}_{K-k}$, given in Equation 2.59:

$$\begin{aligned} \boldsymbol{E}_{K-k} &= \boldsymbol{P}_{K-k} + \boldsymbol{K}_{K-k} \operatorname{cov}(z_k \mid z_k \in A_k) \boldsymbol{K}_{K-k}^T \\ &= \boldsymbol{M}_{K-k} - \boldsymbol{K}_{K-k}(\mathbf{\Gamma}_{K-k} - \operatorname{cov}(z_k \mid z_k \in A_K))\boldsymbol{K}_{K-k}^T, \end{aligned} \tag{2.77}$$

where the latter expression is found by using Equation 2.72. Define the matrix $\boldsymbol{C}_{K-k}$ as

$$\boldsymbol{C}_{K-k} = \mathbf{\Gamma}_{K-k} - \operatorname{cov}(z_k \mid z_k \in A_k). \tag{2.78}$$

Apply the matrix identity Equation 2.75 to Equation 2.77 after substituting Equation 2.78; this yields

$$\begin{aligned} \boldsymbol{E}_{K-k}^{-1} = \boldsymbol{M}_{K-k}^{-1} + \boldsymbol{M}_{K-k}^{-1}\boldsymbol{K}_{K-k}(\boldsymbol{C}_{K-k}^{-1} - \boldsymbol{K}_{K-k}^T\boldsymbol{M}_{K-k}^{-1}\boldsymbol{K}_{K-k})^{-1} \\ \times \boldsymbol{K}_{K-k}^T\boldsymbol{M}_{K-k}^{-1}. \end{aligned} \tag{2.79}$$

The following equalities can be found from Equations 2.69 to 2.71:

$$\begin{aligned} M_{K-k}^{-1} K_{K-k} &= H_k^T \Gamma_{K-k}^{-1}, \\ K_{K-k}^T M_{K-k}^{-1} K_{K-k} &= \Gamma_{K-k}^{-1} F_{K-k} \Gamma_{K-k}^{-1}. \end{aligned} \tag{2.80}$$

These results are used in Equation 2.79 to provide the relation

$$E_{K-k}^{-1} = M_{K-k}^{-1} + H_k^T [\Gamma_{K-k}^{-1}(C_{K-k}^{-1} - \Gamma_{K-k}^{-1} F_{K-k} \Gamma_{K-k}^{-1})^{-1} \Gamma_{K-k}^{-1}] H_k. \tag{2.81}$$

This equation describes the a posteriori information matrix E_{K-k}^{-1} in terms of the a priori information matrix M_{K-k}^{-1} and Γ_{K-k}^{-1}, C_{K-k}^{-1} and F_{K-k}. Because the latter three matrices are functions of M_{K-k}, it appears that M_{K-k}^{-1} must be inverted when its inverse exists. Except for the possibility of numerical difficulties (Fraser, 1967) it would be more convenient to work with E_{K-k} and M_{K-k} directly.

The backward Gaussian fit filter starts with no a priori information at time t_K, so M_{K-1} does not exist. The manner in which the information matrix is updated will be determined by finding the limiting form of the bracketed term in Equation 2.81. For notational convenience let

$$B_k = \operatorname{cov}(z_k \mid z_k \in A_k). \tag{2.82}$$

With this the bracketed term in Equation 2.81 may be written

$$\begin{aligned} [\cdots] &= [\Gamma_{K-k}(\Gamma_{K-k} - B_k)^{-1} \Gamma_{K-k} - F_{K-k}]^{-1} \\ &= [\Gamma_{K-k}(I - \Gamma_{K-k}^{-1} B_k)^{-1} - F_{K-k}]^{-1} \\ &= [\Gamma_{K-k}(I + \Gamma_{K-k}^{-1} B_k + (\Gamma_{K-k}^{-1} B_k)^2 + \cdots) - F_{K-k}]^{-1}. \end{aligned} \tag{2.83}$$

The series expression for the inverse is valid if the elements of Γ_{K-k}^{-1} are decreasing in magnitude as the elements of M_{K-k} increase. Finally, using Equation 2.70, we have

$$\lim_{\Gamma_{K-k}^{-1} \to 0} (\quad) = (R_k + B_k)^{-1} \tag{2.84}$$

and, from Appendix A,

$$\lim_{\Gamma_{K-k}^{-1} \to 0} B_k = \lim_{\Gamma_{K-k}^{-1} \to 0} \operatorname{cov}(z_k \mid z_k \in A_k) = \left(\frac{(\alpha^i)^2}{3} \delta_{ij} \right). \tag{2.85}$$

Substituting Equations 2.84 and 2.85 in Equation 2.81 yields the equation for updating the information matrix when its inverse does not exist:

$$\boldsymbol{E}_{K-k}^{-1} = \boldsymbol{M}_{K-k}^{-1} + \boldsymbol{H}_k^T \left[\boldsymbol{R}_k + \left(\frac{(\alpha^i)^2}{3} \delta_{ij} \right) \right]^{-1} \boldsymbol{H}_k. \tag{2.86}$$

Note that, if the sample lies in a semi-infinite interval, no information is added by this measurement. Hence, Gaussian fit smoothing cannot improve estimates when the measurements have been quantized to two levels.

2.8 Summary

This chapter contains much of the material that is basic to the theory of nonlinear estimation with quantized measurements. First the concept of the maximum-likelihood estimate was introduced. Next it was shown that the expectation of a function of a parameter vector conditioned on quantized measurements can be found in two steps: finding the expectation conditioned on a measurement z and averaging this function of z conditioned on the event $z \in A$. Finally, it was shown that, when applied to discrete-time Gaussian linear systems, the conditioned mean of the system's state vector can be found without Bayes' rule by passing the conditional mean of the measurement history through the Kalman filter that would be used had the measurements been linear; this is true for the continuous-time Gaussian linear system as well. In this chapter different methods of calculating the mean and covariance of the measurement vector were discussed.

Three approximate nonlinear estimation techniques were described. The first is applicable when the quantum intervals are small; it was shown that the filter derived from the quantization-noise model is the minimum variance estimate within the approximation being made. The second technique, the Gaussian fit algorithm, assumes that the conditional distribution is normal; this method may be used with arbitrary quantization schemes and is a recursive calculation. The third technique is a smoothing algorithm that combines the outputs of two Gaussian fit filters.

3 Three Digital Communication Systems

3.1 Introduction

This chapter considers the design of three types of digital communication systems: pulse code modulation (PCM), predictive quantization, and predictive-comparison data-compression. Each of these systems is examined in the light of the approximate nonlinear estimation techniques considered in the last chapter. The Gaussian fit algorithm is especially useful in the two systems employing feedback, because the feedback function can be determined without iteration from the parameters of the input process. Ensemble performance estimates of the Gaussian fit algorithm are derived for each of the three systems, and the accuracy of these estimates is tested by computer simulation.

3.2 Pulse Code Modulation (PCM)

The noiseless-channel version of the problem of pulse-code-modulation is shown in Figure 3.1, where it is assumed that $\{z_n\}$ is a Gaussian process. Note that the quantizer output is the interval A_n in which the scalar sample z_n falls.

When the quantum intervals are small enough, the results on parameter-estimation given in Section 2.4 can be used for the conditional-mean receiver, and the conditional mean of the measurement history is given (approximately) by Equation 2.46. In this equation z is that portion of $\{z_n\}$ upon which the estimate is based. If $\{z_n\}$ is

Figure 3.1 Noiseless-channel pulse-code-modulation system.

derived from a Gauss–Markov process, the alternative estimate based on the quantization-noise model may be solved recursively with the Kalman filter. This filter uses the augmented observation-noise covariance that depends on the intervals in which the measurements lie. For arbitrarily large quantum intervals the Gaussian fit algorithm may be applied in a straightforward manner for approximating the conditional-mean receiver for the filtering and prediction problems. The smoothing problem is more complex; it is briefly considered in Section 2.7. An estimate of the ensemble performance of the Gaussian fit algorithm in the pulse-code-modulation mode is given in Appendix B.

3.3 Predictive Quantization

Figure 3.2 shows the noiseless-channel version of the predictive-quantization problem for a mean-square-error criterion. This system configuration is not as general as the one considered by Fine (1964), since the quantizer is time-invariant. The scalar random process $\{z_n\}$ is assumed to be the output of a system described by Equations 2.28 to 2.33. The N-level quantizer is chosen beforehand but is fixed once the system is in operation. The scalar feedback function $L_n(A_{n-1}, A_{n-2}, \ldots)$ is subtracted from the incoming sample z_n, so as to minimize the mean square reconstruction error. The determination of L_n and the design of the quantizer are considered next.

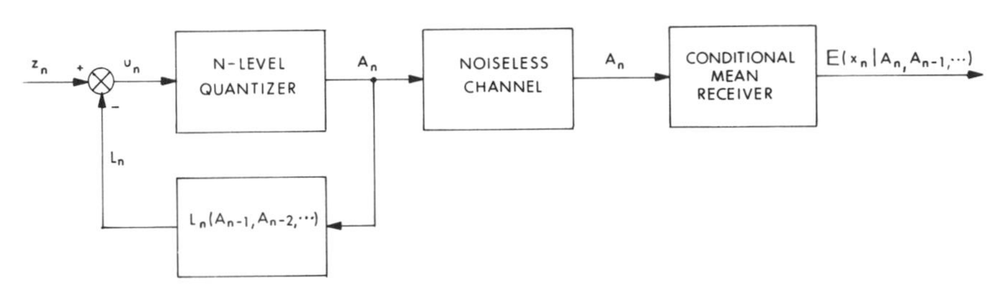

Figure 3.2 Noiseless-channel predictive-quantization system.

System Optimization

Fine (1964) outlines the system design procedure in three steps: finding the optimal receiver for a given transmitter, finding the optimal transmitter for a given receiver, and solving the simultaneous conditions of the first two steps for the optimal system. These are necessary but not sufficient conditions (Fine, 1964). Here we have already performed the first step for the quadratic criterion, since the conditional-mean receiver is indicated in Figure 3.2. The second and third steps are carried out by choosing the optimal feedback quantity L_n^o such that

$$E[\operatorname{cov}(\boldsymbol{x}_n \mid L_n, A_n, A_{n-1}, \ldots) \mid A_{n-1}, \ldots] - E[\operatorname{cov}(\boldsymbol{x}_n \mid L_n^o, A_n, A_{n-1}, \ldots) \mid A_{n-1}, \ldots] \geq 0, \tag{3.1}$$

where $\operatorname{cov}(\boldsymbol{x}_n \mid \ldots)$ is the conditional covariance of the estimate, and the matrix inequality implies that the left-hand side is positive semidefinite. Choosing L_n in this manner assures a minimum variance estimate for any linear combination of the state variables.

An approximate solution to this equation can be found with the aid of the Gaussian fit algorithm. Under this assumption the conditional moments are given in recursive form by Equations 2.56 to 2.61, and the conditional covariance is given by Equation 2.59, except that the index i is replaced with n. Let the quantized variable be u_n (Figure 3.2) and, furthermore, let

$$\begin{aligned} u_n &= z_n - L_n(A_{n-1}, \ldots), \\ u_n^o &= z_n - L_n^o(A_{n-1}, \ldots). \end{aligned} \tag{3.2}$$

Note that in Equation 2.59 we may use $\operatorname{cov}(u_n \mid \ldots)$ in place of $\operatorname{cov}(z_n \mid \ldots)$, since the expectation is conditioned on $\{A_{n-1}, \ldots\}$ so that u_n and z_n differ only by the constant L_n. If the N quantum intervals are denoted by $\{A^j\}$, with $j = 1, \ldots, N$, then substituting Equation 2.59 in Equation 3.1 and simplifying produce the scalar equation

$$\sum_{j=1}^{N} \operatorname{cov}(u_n \mid u_n \in A^j) P(u_n \in A^j) - \sum_{j=1}^{N} \operatorname{cov}(u_n^o \mid u_n^o \in A^j) P(u_n^o \in A^j) \geq 0. \tag{3.3}$$

These quantities are computed by means of Equations 2.39 and 2.41 under the assumption that z_n (hence u_n and u_n^o) are normally distributed. By symmetry arguments it may be concluded that *for a well-designed quantizer* L_n^o should be the (approximate) conditional mean of z_n:

$$L_n^o(A_{n-1}, \ldots) = E(z_n \mid A_{n-1}, \ldots) = \boldsymbol{H}_n \hat{\boldsymbol{x}}_{n|n-1}. \tag{3.4}$$

If, however, the conditional standard deviation of the prediction of z_n is very much smaller than the quantum intervals, then the quantizer appears to be $N - 1$ binary quantizers placed end to end. In this case L_n^o will be so chosen as to place the mean of u_n^o at one of these quantizer switch points. The quantizer design problem is considered in more detail in Appendix B.

It will be assumed that the quantum intervals are small enough for the optimal choice of L_n to be the conditional mean of z_n based on measurements at $t_{n-1}, t_{n-2}, \ldots$. The feedback function L_n may be based on a limited memory or a growing memory. The feedback function for the growing-memory version uses the Gaussian fit algorithm on stationary or nonstationary processes from the first measurement to the most recent. Although the *memory* is increasing, the *storage* required to implement the equations remains constant. The great advantages of the Gaussian fit algorithm are that its operation is determined by the parameters of the input process and that it is not an iteratively determined function, as are most feedback operations of this type. The ensemble mean square error of this system is described in Appendix B.

There is one special case in which the operation of the Gaussian fit algorithm becomes stationary and the feedback function and receiver become linear. When a binary quantizer is used, the filter weights as given by Equations 2.57, 2.58, 2.59, and 2.61 are predetermined functions of time, because $\operatorname{cov}(z_i \mid \ldots)$ does not depend on the measurements. In a stationary input process the filter weights will approach a constant as more measurements are incorporated.

From Equations 2.39, 2.56, 2.57, 2.60, and 3.4 the state-space form of the stationary equations is

$$\hat{\boldsymbol{x}}_{n|n-1} = \boldsymbol{\Phi} \hat{\boldsymbol{x}}_{n-1|n-1}, \tag{3.5}$$

$$L_n^o = \boldsymbol{H} \hat{\boldsymbol{x}}_{n|n-1}, \tag{3.6}$$

$$\hat{\boldsymbol{x}}_{n|n} = \hat{\boldsymbol{x}}_{n|n-1} + \boldsymbol{K}_\infty \hat{r}_n, \tag{3.7}$$

$$\hat{r}_n = \sqrt{2/\pi}\,\sigma_{z_\infty} \operatorname{sgn}(z_n - L_n^o), \tag{3.8}$$

$$\sigma_{z_\infty}^2 = \boldsymbol{H} \boldsymbol{M}_\infty \boldsymbol{H}^T + \boldsymbol{R}, \tag{3.9}$$

$$\boldsymbol{K}_\infty = \boldsymbol{M}_\infty \boldsymbol{H}^T (\boldsymbol{H} \boldsymbol{M}_\infty \boldsymbol{H}^T + \boldsymbol{R})^{-1}, \tag{3.10}$$

where M_∞ is the steady-state solution to the covariance equations, Equations 2.58, 2.59, and 2.61.

3.4 Predictive-Comparison Data-Compression

Figure 3.3 is a block diagram of the predictive-comparison type of data-compression system. The analysis contained here is concerned only with the prediction and filtering aspects; such important problems as buffer-control, timing information, and channel noise are not considered.

The threshold device in Figure 3.3 is a quantizer (one large quantum interval, many small quantum intervals), and the linear slope indicates that quantization during encoding may be neglected. The quantizer output is fed back through $L_n(A_{n-1}, \ldots)$ and subtracted from the input z_n. If the magnitude of the difference, $|u_n|$, is less than α (a known parameter), then nothing is sent to the receiver; if $|u_n|$ is greater than α, then u_n is sent to the receiver. The receiver does a parallel computation of $L_n(A_{n-1}, \ldots)$, and the system input z_n is calculated by adding u_n and L_n when u_n is sent. The estimate of z_n when u_n is *not* sent may be made according to a variety of criteria; in all cases the error in the estimate is known to be less than α.

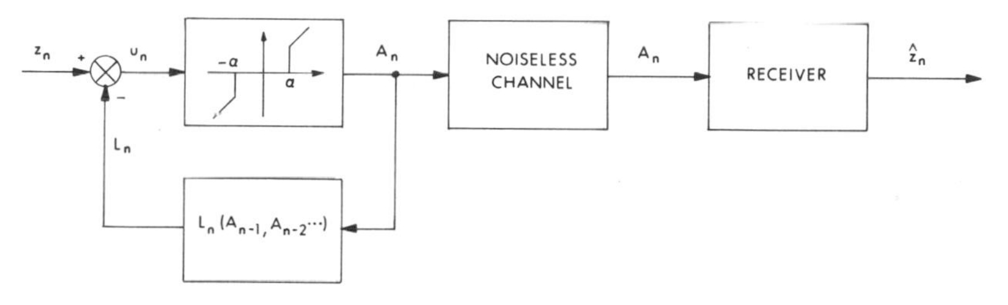

Figure 3.3 Noiseless-channel predictive-comparison data-compression system.

System Optimization

The criterion for the optimal system will not contain data fidelity, since this can be controlled through the choice of the threshold width. Instead, $L_n(A_{n-1}, \ldots)$ is chosen solely on the basis that it shall minimize the conditional probability that u_n is sent to the receiver. This is not necessarily the same as minimizing the average number of samples sent out of the total number processed; the latter problem may be formulated as one in optimal stochastic control requiring a dynamic-programming approach.

The necessary equations for optimality are treated next. Let $y_{n-1} = \{z_{n-1}, z_{n-2}, z_{n-3}, \ldots\}$ be the measurements used in determining L_n, and let $\tilde{A}_{n-1} = \{A_{n-1}, A_{n-2}, \ldots\}$ be the region in which they fall. Then the probability of rejecting (not sending) the nth sample is

$$P(\text{reject}) = \int_{L_n - \alpha}^{L_n + \alpha} \underset{z_n | y_{n-1} \in \tilde{A}_{n-1}}{p(\zeta_n)} \, d\zeta_n \tag{3.11}$$

where

$$\underset{z_n | y_{n-1} \in \tilde{A}_{n-1}}{p(\zeta_n)} = \frac{\displaystyle\int_{\tilde{A}_{n-1}} \underset{z_n, y_{n-1}}{p(\zeta_n, \eta_{n-1})} \, d\eta_{n-1}}{\displaystyle\int_{\tilde{A}_{n-1}} \underset{y_{n-1}}{p(\eta_{n-1})} \, d\eta_{n-1}}. \tag{3.12}$$

The necessary condition that the probability of rejection be stationary with respect to L_n at L_n^o is

$$\underset{z_n | y_{n-1} \in \tilde{A}_{n-1}}{p(L_n^o + \alpha)} = \underset{z_n | y_{n-1} \in \tilde{A}_{n-1}}{p(L_n^o - \alpha)} \tag{3.13}$$

or, on substitution of Equation 3.12 in Equation 3.13,

$$\int_{\tilde{A}_{n-1}} \underset{z_n, y_{n-1}}{p(L_n^o + \alpha, \eta_{n-1})} \, d\eta_{n-1} = \int_{\tilde{A}_{n-1}} \underset{z_n, y_{n-1}}{p(L_n^o - \alpha, \eta_{n-1})} \, d\eta_{n-1}. \tag{3.14}$$

This expression does not require that $\{z_n\}$ be derived from a Markov process. If the threshold halfwidth α is small enough, Equation 3.13 can be solved by power-series approximations. Neglecting terms of third and higher order, it may be verified that Equation 3.13 reduces to

$$\frac{\partial}{\partial \zeta_n} \left(\underset{z_n, y_{n-1}}{p(\zeta_n, \gamma_{n-1})} \right) \Bigg|_{L_n^o} = 0, \tag{3.15}$$

where γ_{n-1} is the collection of midpoints of the quantum intervals $A_{n-1}, A_{n-2}, \ldots$. Note that this equation is equivalent to

$$\frac{\partial}{\partial \zeta_n} \left(\underset{z_n | y_{n-1}}{p(\zeta_n, \gamma_{n-1})} \right) \Bigg|_{L_n^o} = 0, \tag{3.16}$$

which means that to the second-order approximation the optimal L_n is the mode of the density function of z_n conditioned on unquantized measurements γ_{n-1}. If the z_n are Gaussian random variables, then the conditional mode is the conditional mean, and L_n^o is a linear operation.

When $\{z_n\}$ is derived from a Gauss–Markov process, the Gaussian fit algorithm may be used in the feedback path for arbitrarily wide thresholds. Regardless of the number of samples that have been rejected or sent, the distribution of z_n conditioned on quantized measurements at times $t_{n-1}, \ldots$, is assumed to be normal. The conditional probability of rejecting z_n is maximized by choosing a value of L_n that is the conditional mean (approximately), $\boldsymbol{H}_n\hat{\boldsymbol{x}}_{n|n-1}$. The feedback function $L_n^o = E(z_n \mid A_{n-1}, \ldots)$ is computed with the Gaussian fit algorithm. Observe that $E(z_n \mid z_n \in A_n)$ is just L_n if the sample falls within the threshold and is z_n if the sample is not quantized. The ensemble performance estimate of the system made with the Gaussian fit algorithm is derived in Appendix B.

Alternative approaches to system design rely on feeding back the prediction (in some sense) of the next sample (say) z_n (Irwin and O'Neal, 1968; Davisson, 1967; Ehrman, 1967). Davisson (1967) considers a stationary Gaussian input sequence and small threshold widths; he finds the optimal linear predictor for z_n based on the most recent M samples and uses it in the feedback path. By drawing on the results of Sections 2.4 and 2.7 the optimal nonlinear predictor is found from Equation 2.22, in which $\boldsymbol{x}$ is replaced with z_n and z with $\{z_{n-1}, \ldots, z_{n-M}\}$ and $E(z \mid z \in A)$ is given by Equation 2.46. The net result is a set of filter weights for $\{\gamma_{n-1}, \ldots, \gamma_{n-M}\}$, each of which takes on one of 2^M values, depending on which samples have been quantized.

3.5 System Simulations

This section describes the results of digital computer simulations of the Gaussian fit algorithm as applied to the pulse-code-modulation, predictive-quantization, and data-compression systems. Bello and his associates (1967) present simulation results for predictive quantization with a binary quantizer. Their approach is a numerical approximation (by Monte Carlo techniques) to the optimal feedback function, whereas ours is an analytical approximation (the Gaussian fit algorithm). They consider various memory lengths and a binary quantizer, and here we use a growing memory (finite storage) and arbitrary quantizers. Although the Gaussian fit algorithm and its performance estimate may be used on nonstationary data, only stationary data have been simulated as yet.

Simulation Description

Input Process The second-order, Gauss–Markov input process is the sampled output of a linear system driven by Gaussian white noise.

The transfer function of the shaping filter is the same as that used by Bello and his co-workers,

$$H(s) = \frac{c}{(1 + \tau s)^2}, \tag{3.17}$$

where the gain c is so chosen as to provide the proper variance at the output. Observation noise is not used here, but it is used in Chapter 4. Thus the autocorrelation of the input process is

$$\phi_{zz}(n) = E(z_i z_{i+n}) = (1 + |n|/r) \exp(-|n|/r), \tag{3.18}$$

where $r = \tau/T$ is the number of samples per time constant τ and T is the time between samples.

Error Measurement Each system was simulated by operating on five thousand consecutive samples. The estimation errors were squared and averaged to give an estimate of the ensemble mean square error of the system. The autocorrelations of the estimation errors were measured, and from this the confidence limits were assessed as being greater than a 90 percent probability that the measured error variance lies within 10 percent of its true value.

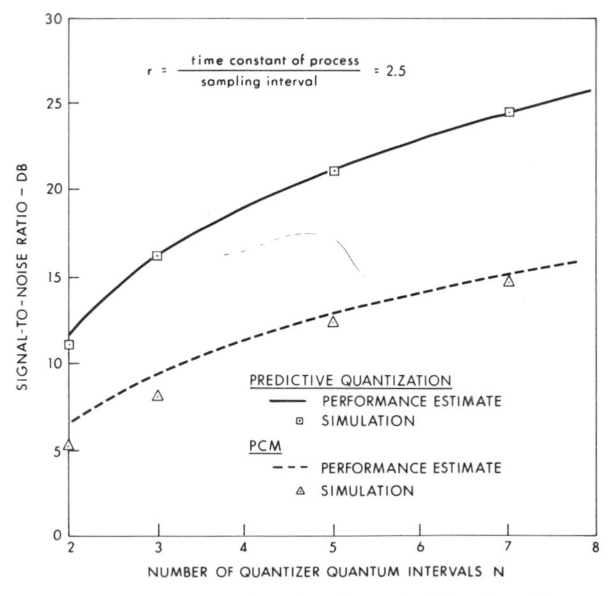

Figure 3.4 Signal-to-noise ratio for the Gaussian fit algorithm: pulse code modulation and predictive quantization.

PCM and Predictive Quantization

Figure 3.4 displays the ratio of signal variance to ensemble mean square estimation error (expressed in decibels) as a function of the number of quantizer quantum intervals. The figure represents both the pulse-code-modulation and predictive-quantization systems with the input process parameter $r = 2.5$. The performance estimates are as derived in Appendix B. The predictive-quantization system performs significantly better than the PCM system, as is to be expected. The performance estimate is quite accurate except for PCM with a small number of quantum intervals (less than 5). Here the estimate is optimistic, a characteristic that has been noted in other simulations (Curry, 1968). The quantizer quantum intervals have been so chosen as to minimize the ensemble mean square error (see Appendix B).

Figure 3.5 shows how the predictive-quantization system with a binary quantizer reacts to different signal correlations. The performance estimate and the simulation results are exhibited as a function

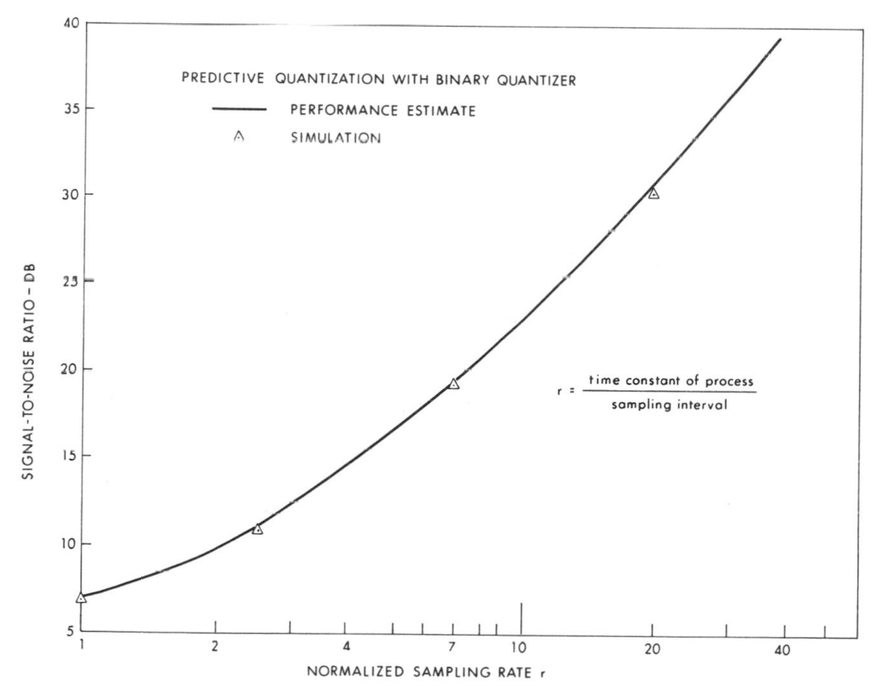

Figure 3.5 Signal-to-noise for the Gaussian fit algorithm: predictive quantization with binary quantizer.

of the input process parameter r (as a point of reference, the adjacent sample correlation is 0.736 for $r = 1$, 0.938 for $r = 2.5$, and 0.9988 for $r = 20$). Again the performance estimate is quite accurate.

Data Compression

Figure 3.6 contains the outcomes for the predictive-comparison data-compression system. Performance estimates and simulation results of the mean square error and sample-compression ratio* are shown as a function of α/σ_z^a, the ratio of threshold halfwidth to a priori standard deviation. Note the excellent agreement between performance estimates and simulation results.

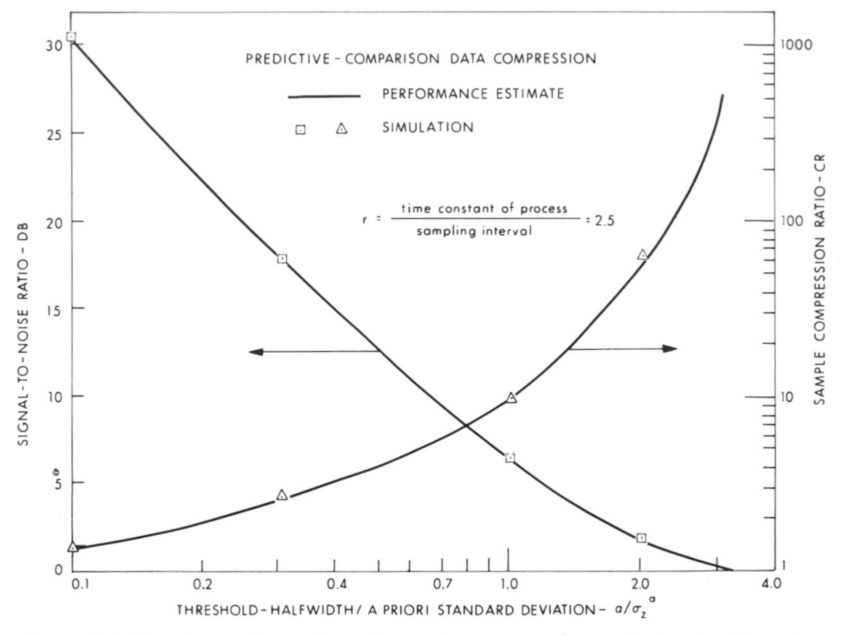

Figure 3.6 Signal-to-noise ratio and sample-compression ratio for the Gaussian fit algorithm: predictive-comparison data-compression.

3.6 Summary

Approximate nonlinear estimation schemes were applied to the noiseless-channel versions of three digital systems: pulse code modula-

* The sample-compression ratio is the number of input samples divided by the average number of samples transmitted to the receiver.

tion, predictive quantization, and predictive-comparison data-compression. These methods can be used on stationary and nonstationary data and can be used in the feedback path without additional calculations, such as Monte Carlo methods. The Gaussian fit algorithm uses a growing memory (but finite storage) for these computations. Estimates of the ensemble mean square reconstruction error are derived for the Gaussian fit algorithm when used in each of the three systems. Simulation results indicate that these ensemble performance estimates are quite accurate (except for very coarse pulse code modulation), so that parametric studies with Monte Carlo techniques are not required for evaluating the system's ensemble mean square error.

4 Optimal Linear Estimators for Quantized Stationary Processes

4.1 Introduction

This chapter examines the calculation of minimum variance linear estimators when the measurements are quantized, stationary random processes. This class of filters is important, because the linear filter is relatively easy to implement and a stationary process is an adequate model in many situations. Consideration is limited to linear discrete-time filters. The autocorrelation function of the quantizer output is examined in some detail because of its importance for finding the optimal linear filter. The quantizer is modeled as a gain element and an additive-noise source. New criteria are proposed for choosing this gain element, and these lead to new interpretations of the random-input describing function. Computation of the minimum variance linear filter is discussed, and the performance is compared with that given by the Gaussian fit algorithm. The last portion of the chapter treats the joint optimization of quantizer and filter for investigating improvements in performance that might be available with this additional freedom.

4.2 Autocorrelation of the Quantizer Output

Throughout this chapter it will be assumed that the quantizer has a fixed input-output relationship; three examples are shown in Figure 4.1. Whenever the input sample falls in the quantum interval A^i, whose lower and upper limits are d^i and d^{i+1}, respectively, then the quantizer

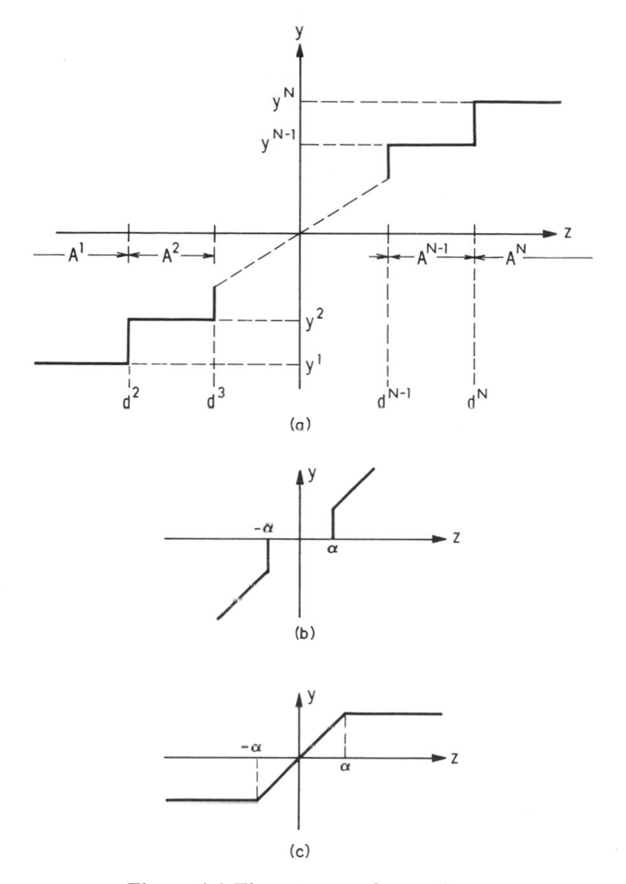

Figure 4.1 Three types of quantizers.

output is y^i. To be consistent with the equations that follow, the linear slopes in Figures 4.1b and 4.1c may be considered regions of infinitesimally small quantum intervals.

The autocorrelation function for the quantizer output must be calculated from the second-order probability-density function of the zero-mean, wide-sense stationary, scalar-valued z process:

$$\phi_{yy}(m) = E(y_n y_{n+m}) = \sum_{i=1}^{N} \sum_{j=1}^{N} y^i y^j P(z_n \in A^i, z_{n+m} \in A^j), \tag{4.1}$$

where

$$P(z_n \in A^i, z_{n+m} \in A^j) = \int_{d^i}^{d^{i+1}} d\zeta_1 \int_{d^j}^{d^{j+1}} d\zeta_2 \, p_{z_n, z_{n+m}}(\zeta_1, \zeta_2), \tag{4.2}$$

$$
\begin{aligned}
z_n &= \text{quantizer input at time } t_n, \\
y_n &= \text{quantizer output at time } t_n, \\
A^i &= i\text{th quantum interval, with } i = 1, 2, \ldots, N, \\
y^i &= \text{quantizer output for input in } A^i, \\
d^i &= \text{lower limit of } A^i, \\
\underset{z_n, z_{n+m}}{p}(\zeta_1, \zeta_2) &= \text{joint probability-density function of } z_n \text{ and } z_{n+m}.
\end{aligned}
$$

For a stationary Gaussian process the explicit form of the joint probability, Equation 4.2, is

$$
\begin{aligned}
P(z_n \in A^i, z_{n+m} \in A^j) = \int_{d^i}^{d^{i+1}} & \frac{\exp[-\frac{1}{2}(\zeta_1^2/\sigma_z^2)]}{(2\pi)^{1/2}\sigma_z} \\
& \times \left[F\left(\frac{d^{j+1} - \rho_z(m)\zeta_1}{\sigma_z[1 - \rho_z^2(m)]^{1/2}} \right) \right. \\
& \left. - F\left(\frac{d^j - \rho_z(m)\zeta_1}{\sigma_z[1 - \rho_z^2(m)]^{1/2}} \right) \right] d\zeta_1, \qquad (4.3)
\end{aligned}
$$

where

$$F(x) = \int_{-\infty}^{x} \frac{\exp(-\frac{1}{2}u^2)}{(2\pi)^{1/2}} \, du, \qquad (4.4)$$

$$\rho_z(m) = \frac{E(z_n z_{n+m})}{\sigma_z^2}. \qquad (4.5)$$

Note that the correlation between two input samples determines the correlation between the same two output samples through Equations 4.3 and 4.1 and the fact that

$$\rho_y(m) = \frac{\phi_{yy}(m)}{\overline{y^2}}, \qquad (4.6)$$

where

$$\overline{y^2} = \sum_{i=1}^{N} (y^i)^2 P(z \in A^i), \qquad (4.7)$$

$$P(z \in A^i) = F\left(\frac{d^{i+1}}{\sigma_z}\right) - F\left(\frac{d^i}{\sigma_z}\right). \qquad (4.8)$$

The Quantizer: A Gain and Additive Noise

The quantizer may be replaced with a gain and an additive-noise source, the noise being defined as the difference between the quantizer output and the gain output. In general, this noise will be correlated with the quantizer input. The purpose of this approach is to examine the characteristics of the noise source and to determine under what conditions the effects of quantization may be neglected.

Figure 4.2 shows the quantization operation Q replaced with a gain G and additive noise n_G. The quantizer input and output satisfy the equations

$$y = Q(z), \tag{4.9}$$

$$y = Gz + n_G. \tag{4.10}$$

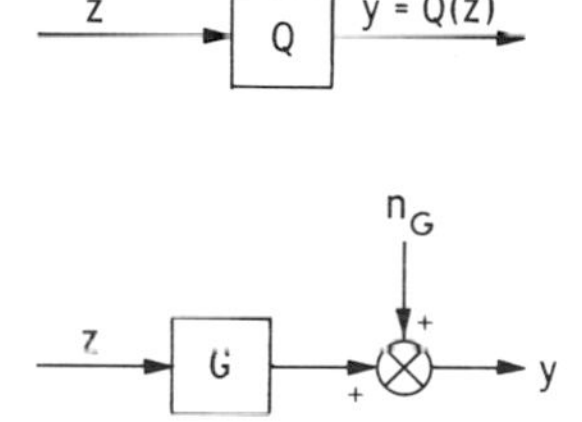

Figure 4.2 The quantizer as a gain and additive noise.

More precisely, the noise is defined by Equation 4.10 and will have different characteristics for different values of G:

$$n_G \equiv y - Gz. \tag{4.11}$$

If the noise process were known, then the autocorrelation function of the quantizer output could be calculated as follows:

$$\phi_{yy}(m) = G^2\phi_{zz}(m) + G\phi_{zn_G}(m) + G\phi_{n_Gz}(m) + \phi_{n_Gn_G}(m). \tag{4.12}$$

But, from Equation 4.11,

$$\begin{aligned} \phi_{n_Gz}(m) &= \phi_{yz}(m) - G\phi_{zz}(m), \\ \phi_{zn_G}(m) &= \phi_{zy}(m) - G\phi_{zz}(m). \end{aligned} \tag{4.13}$$

These results are substituted in Equation 4.12 and yield

$$\phi_{yy}(m) = -G^2\phi_{zz}(m) + G[\phi_{zy}(m) + \phi_{yz}(m)] + \phi_{n_Gn_G}(m). \tag{4.14}$$

The remainder of this section is devoted to zero-mean Gaussian processes. The cross correlation between the output y and input z for an arbitrary zero-memory nonlinearity $y = f(z)$ is found as follows:

$$\phi_{yz}(m) = E(y_n z_{n+m}) = \int_{-\infty}^{\infty} d\zeta_n f(\zeta_n) p_{z_n}(\zeta_n)$$

$$\times \int_{-\infty}^{\infty} d\zeta_{n+m} \zeta_{n+m} \underset{z_{n+m}|z_n}{p}(\zeta_{n+m}, \zeta_n)$$

$$= \int_{-\infty}^{\infty} d\zeta_n f(\zeta_n) p_{z_n}(\zeta_n) E(z_{n+m} \mid z_n = \zeta_n). \tag{4.15}$$

But for Gaussian random variables it is true that

$$E(z_{n+m} \mid z_n = \zeta_n) = \frac{\phi_{zz}(m)}{\sigma_z^2} \zeta_n,$$

and Equation 4.15 can then be written

$$\phi_{yz}(m) = \frac{\phi_{zz}(m)}{\sigma_z^2} \int_{-\infty}^{\infty} d\zeta_n \zeta_n f(\zeta_n) p_{z_n}(\zeta_n),$$

$$= \mathsf{k} \phi_{zz}(m) \tag{4.16}$$

where, in general,

$$\mathsf{k} = \frac{E(y_n z_n)}{\sigma_z^2} = \frac{1}{\sigma_z^2} \int_{-\infty}^{\infty} dz\, z\, f(z)\, p(z)$$

and, for quantizers,

$$\mathsf{k} = \frac{1}{\sigma_z^2} \sum_{i=1}^{N} y^i \int_{d^i}^{d^{i+1}} dz\, z\, p(z). \tag{4.17}$$

The quantity k is the describing-function gain (Gelb and Vander Velde, 1968); it is the linear approximation to the nonlinearity that minimizes the mean square approximation error. Now substitute Equation 4.16 in Equation 4.14 to obtain an expression for the output correlation ϕ_{yy}:

$$\phi_{yy}(m) = G(2\mathsf{k} - G)\, \phi_{zz}(m) + \phi_{n_G n_G}(m). \tag{4.18}$$

Thus, regardless of the gain G that is used, the quantizer output correlation may be computed as though the quantizer output came from two sources: a signal source, which is $[G(2\mathsf{k} - G)]^{1/2}$ times the quantizer input, and an additive, uncorrelated noise source n_G.

4.3 A New Interpretation of the Describing Function

The difficulty in analysis and synthesis arises when the output autocorrelation function needs to be evaluated. Many times a quantizer is modeled as a unit gain ($G = 1$) and a white noise source (Widrow, 1960; Ruchkin, 1961; Steiglitz, 1966; O'Neal, 1966; Irwin and O'Neal, 1968). Widrow (1960) neglects the input-noise cross correlation as well. It is well known, however, that choosing the value of G to be the describing-function gain k will minimize the variance of the noise source, $\phi_{n_G n_G}(0)$. The gain may be chosen on the basis of other considerations, and some of them are suggested below.

For example, in the analysis process it is convenient to have the autocorrelation of the noise source be of little significance in determining the output autocorrelation; that is, we should like to minimize by our choice of G the ratio

$$|\phi_{n_G n_G}(m)/\phi_{yy}(m)|. \tag{4.19}$$

It will be shown for zero-mean, Gaussian, random processes and odd nonlinearities that $G = \mathrm{k}$ provides a global minimum of this ratio for all values of the time argument m. Since ϕ_{yy} is not influenced by the choice of G, this value of G also minimizes $|\phi_{n_G n_G}(m)|$. This criterion may be thought of as a whiteness property, and the "whitest" noise source may provide simplifications in synthesis and analysis procedures. Lastly, in the synthesis process it is convenient to think of the quantizer output as the sum of a signal source and an uncorrelated noise source. It will be shown that when G is the describing function, the "noise-to-signal ratio,"

$$\left|\frac{\phi_{n_G n_G}(m)}{G(2\mathrm{k} - G)\,\phi_{zz}(m)}\right| \tag{4.20}$$

is at a local minimum (the other extrema occur at $G = \pm\infty$). This means that the noise source is as white as possible relative to the signal source. These three results depend on the following lemma.

Lemma For zero-mean Gaussian processes and odd, zero-memory nonlinearities it is true that

$$|\phi_{yy}(\tau)| \geq \mathrm{k}^2\phi_{zz}(\tau), \qquad 0 \leq \tau \leq \infty. \tag{4.21}$$

Proof Expand the joint normal probability-density function for $z(t)$ and $z(t + \tau)$ in Hermite polynomials. Then $\phi_{yy}(\tau)$ may be written

(Gelb and Vander Velde, 1968, Appendix E):

$$\phi_{yy}(\tau) = \sum_{m(\text{odd})=1}^{\infty} a_m^2 \rho_z^m(\tau)$$

$$= \mathsf{k}^2 \phi_{zz}(\tau) + \sum_{m(\text{odd})=3}^{\infty} a_m^2 \rho_z^m(\tau), \tag{4.22}$$

where $\rho_z(\tau)$ is the correlation coefficient between $z(t)$ and $z(t + \tau)$, and

$$a_m = \frac{1}{(2\pi m!)^{1/2}} \int_{-\infty}^{\infty} y(\sigma_z u) \exp\left(-\frac{u^2}{2}\right) H_m(u)\, du, \tag{4.23}$$

$$H_m(u) = (-1)^m \exp\left(\frac{u^2}{2}\right) \frac{d^m}{du^m}\left[\exp\left(-\frac{u^2}{2}\right)\right]. \tag{4.24}$$

The assertion of the lemma follows from the fact that ϕ_{zz} is an odd function of $\rho_z(\tau)$ and that the sum on the right-hand side of Equation 4.22 is positive for $\rho_z > 0$ and negative for $\rho_z < 0$.

To prove that the describing function minimizes the significance of the ratio of noise autocorrelation to output autocorrelation we use Equation 4.18 to write

$$\left|\frac{\phi_{n_G n_G}(m)}{\phi_{yy}(m)}\right| = \left|1 - G(2\mathsf{k} - G)\frac{\phi_{zz}(m)}{\phi_{yy}(m)}\right|. \tag{4.25}$$

Since ϕ_{zz}/ϕ_{yy} is positive, the global minimum occurs either when $G(2\mathsf{k} - G)$ is at its maximum (which is k^2 at $G = \mathsf{k}$) or when G is such that the second term cancels the first term. The latter possibility is ruled out by the lemma, so that $G = \mathsf{k}$ provides the global minimum. The value of $|\phi_{n_G n_G}(m)|$ is also minimized by this choice of G, since ϕ_{yy} is independent of G.

The noise-to-signal ratio, or criterion of whiteness relative to the signal, is given by

$$\left|\frac{\phi_{n_G n_G}(m)}{G(2\mathsf{k} - G)\,\phi_{zz}(m)}\right| = \left|\frac{\phi_{yy}(m)}{G(2\mathsf{k} - G)\,\phi_{zz}(m)} - 1\right|. \tag{4.26}$$

The argument of the magnitude operation can never be zero, according to the lemma, so it follows that the extremal points occur at $G = \mathsf{k}$ and $G = \pm\infty$. We shall use the local minimum at $G = \mathsf{k}$, which will be a global minimum if $|\phi_{yy}|$ is less than $2\mathsf{k}^2|\phi_{zz}|$.

The noise-to-signal ratio, Expression 4.20, can be calculated as a function of the magnitude of the correlation between two input samples for Gaussian random variables (both ϕ_{yy} and ϕ_{zz} are odd functions of the input correlation, and by Expression 4.20 the noise-to-signal ratio is an even function of the correlation). Figures 4.3 and 4.4 show results for the optimal N-level Gaussian quantizer and the threshold quantizer, respectively. To give some idea of the advantage of using the describing-function gain k rather than unit gain in the quantizer model, Figure 4.3 shows the noise-to-signal ratios for both.

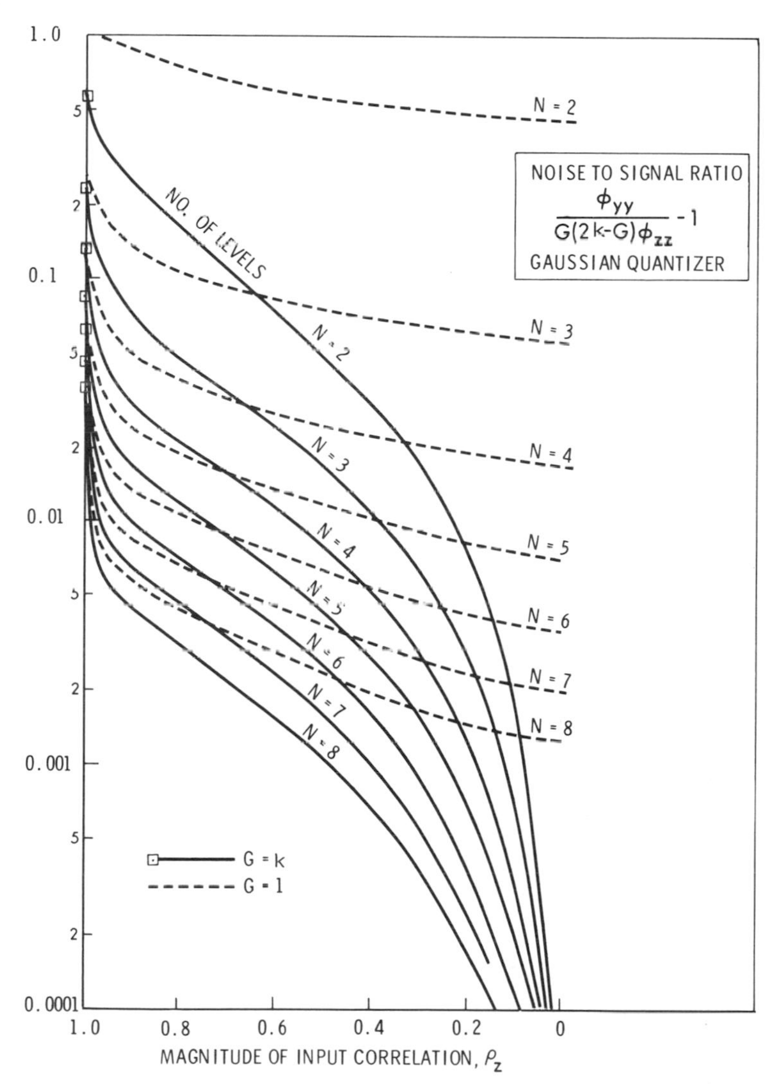

Figure 4.3 Noise-to-signal ratio $\phi_{yy}/G(2\text{k} - G)\phi_{zz} - 1$, Gaussian quantizer.

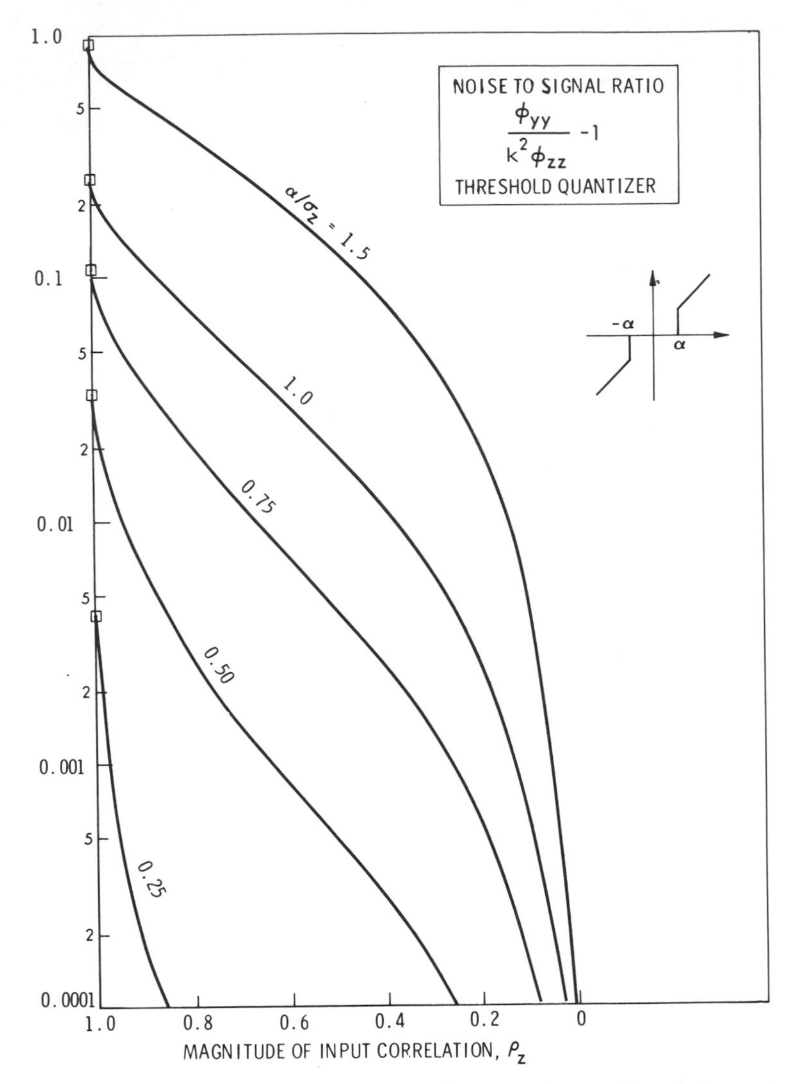

Figure 4.4 Noise-to-signal ratio $\phi_{yy}/k^2\phi_{zz} - 1$, threshold quantizer.

For unit gain, not only is the ratio larger, but it does not go to zero as input correlation goes to zero.

The approximate correlation time for the quantization noise can be determined from these plots: given the autocorrelation function of the zero-mean Gaussian process that is being quantized, the noise-to-signal ratio and, thus, the noise autocorrelation may be determined for each value of input autocorrelation function.

The plots may also be used to determine under what conditions the quantization noise is white. For example, suppose that noise-to-signal ratios of 0.01 or less are considered negligible and that the process is being quantized to two levels. When the quantizer is represented by the describing-function gain k, the quantization noise may be considered white if the correlation between adjacent input samples is less than 0.24. If unit gain is used, the quantization noise may never be considered white.

4.4 Optimal Linear Filters for Quantized Measurements

In this section we briefly consider the problem of designing the optimal linear filter for quantized measurements of a stationary process. Figure 4.5 shows the system to be evaluated: x is the scalar-valued signal, v is white observation noise, y is the quantizer output, and e is the estimation error. The linear filter weights the samples to provide the best linear estimate of the signal x.

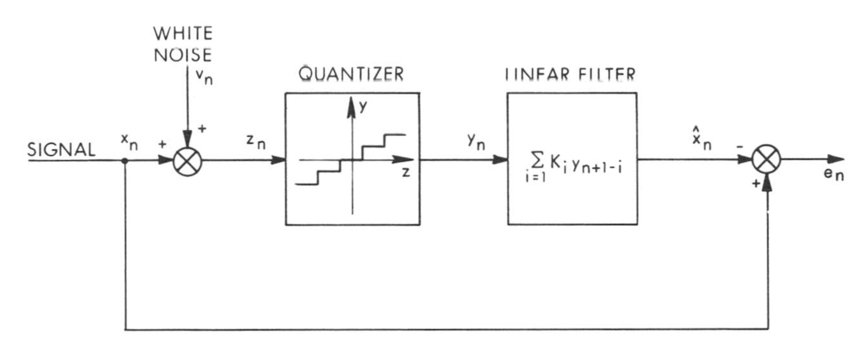

Figure 4.5 Linear filtering problem.

Estimation with a Finite Number of Measurements

When a finite number of measurements are used to estimate the signal matrix methods may be used (Kellog, 1967). Suppose that the value of x_n based on M quantizer outputs $y_n, y_{n-1}, \ldots, y_{n-M+1}$ is to be estimated. Let $\boldsymbol{\eta}_n$ be a vector representing these measurements:

$$\boldsymbol{\eta}_n = \begin{pmatrix} y_n \\ y_{n-1} \\ \vdots \\ y_{n-M+1} \end{pmatrix}, \tag{4.27}$$

and let $\hat{x}_n$ be the minimum variance linear estimate of x_n (without loss of generality it will be assumed that all random variables have zero mean). Then

$$\hat{x}_n = \boldsymbol{K}\boldsymbol{\eta}_n, \tag{4.28}$$

and the choice of the row vector $\boldsymbol{K}$ that minimizes the variance of the estimate (Schweppe, 1967) is

$$\boldsymbol{K} = E(x_n\boldsymbol{\eta}_n^T)\,[E(\boldsymbol{\eta}_n\boldsymbol{\eta}_n^T)]^{-1}. \tag{4.29}$$

The variance in the estimate is given by the following relation (Schweppe, 1967):

$$\overline{e^2} = E(x - \hat{x})^2 = \overline{x^2} - \boldsymbol{K}E(\boldsymbol{\eta}_n\boldsymbol{\eta}_n^T)\boldsymbol{K}^T. \tag{4.30}$$

Because the processes are stationary, the filter gains $\boldsymbol{K}$ may be expressed in terms of the quantities introduced in the previous section. There are only M independent elements in the $M \times M$ covariance matrix of $\boldsymbol{\eta}$, since the ijth element is

$$[E(\boldsymbol{\eta}_n\boldsymbol{\eta}_n^T)]_{ij} = E(y_{n+1-i}y_{n+1-j}) = \phi_{yy}(i - j). \tag{4.31}$$

Furthermore, the ith element of the row vector $E(x_n\boldsymbol{\eta}_n^T)$ is

$$\begin{aligned} [E(x_n\boldsymbol{\eta}_n^T)]_i &= E(x_n y_{n+1-i}) \\ &= \sum_{j=1}^{N} E(x_n \mid z_{n+1-i} \in A^j) y^j P(z_{n+1-i} \in A^j). \end{aligned} \tag{4.32}$$

For Gaussian processes the expectation of x_n conditioned on $z_{n+1-i} \in A^j$, which may be found from the methods of Section 2.4, is

$$\begin{aligned} E(x_n \mid z_{n+1-i} \in A^j) &= \frac{E(x_n z_{n+1-i})}{\sigma_z^2} E(z_{n+1-i} \mid z_{n+1-i} \in A^j) \\ &= \frac{\phi_{xx}(i-1)}{\sigma_z^2} E(z_{n+1-i} \mid z_{n+1-i} \in A^j), \end{aligned} \tag{4.33}$$

where $\phi_{xx}(\ldots)$ is the autocorrelation function of the x process. Substituting this in Equation 4.32 yields*

$$[E(x_n \boldsymbol{\eta}_n^T)]_i = \mathsf{k}\, \phi_{xx}(i - 1), \tag{4.34}$$

where k is the quantizer describing-function gain of Section 4.2:

$$\mathsf{k} = \frac{1}{\sigma_z^2} \sum_{j=1}^{N} E(z \mid z \in A^j) y^j P(z \in A^j). \tag{4.35}$$

The equation of the filter weights, Equation 4.29, may now be expressed as

$$\boldsymbol{K} = \mathsf{k} \boldsymbol{\phi}_{xx}^T \boldsymbol{Y}^{-1}, \tag{4.36}$$

where the M-component vector $\boldsymbol{\phi}_{xx}$ and $M \times M$ matrix $\boldsymbol{Y}$ are defined by

$$\boldsymbol{\phi}_{xx} = \begin{pmatrix} \phi_{xx}(0) \\ \cdot \\ \cdot \\ \phi_{xx}(M - 1) \end{pmatrix}, \tag{4.37}$$

$$\boldsymbol{Y} = \{\phi_{yy}(i - j)\}. \tag{4.38}$$

Estimation with an Infinite Number of Measurements

When the number of measurements used in the estimate becomes infinite, spectrum-factorization techniques must be used to find the filter weights (Raggazzini and Franklin, 1958). The optimal filter will not be a rational function of the Z-transform variable, even if the x process has a rational spectrum, because the spectrum of the quantizer output is not rational. This implies that an infinite number of filter weights must be stored, unless the quantization-noise spectrum is approximated by a rational function (Lanning and Battin, 1956).

An alternative approach is to compute the mean square estimation error for different memory lengths and truncate the memory when the improvement in mean square error with more measurements is negligible.†

The matrix inversion that is required for the computation of the mean square error (Equation 4.30) may be done quite readily with existing computer routines, or it may be done recursively. The recursive computation is performed by partitioning the $(q + 1) \times (q + 1)$

* Smoothing and prediction are easily done by shifting the argument of ϕ_{xx} in Equation 4.34.

† See Grenander and Szego (1958) for a discussion of rates of convergence.

matrix $\boldsymbol{Y}$ into a 1×1 matrix, two vectors, and the $q \times q$ matrix $\boldsymbol{Y}$. The inverse of the $(q + 1) \times (q + 1)$ matrix $\boldsymbol{Y}$ may be found in terms of the inverse of the $q \times q$ matrix $\boldsymbol{Y}$ (Schweppe, 1967). This process may be initiated by calculating the inverse of the 1×1 matrix $\boldsymbol{Y}$ and proceeding to higher dimensions.

Figure 4.6 shows how the mean square estimation error varies with memory length for observations that have been quantized to two levels by the optimal Gaussian quantizer. The computations were carried out for a first-order and second-order Gauss–Markov process. The exponential envelope of the autocorrelation for the second-order process is the same as the autocorrelation function of the first-order process. Note how little improvement there is beyond a memory length of 6 to 8 samples.

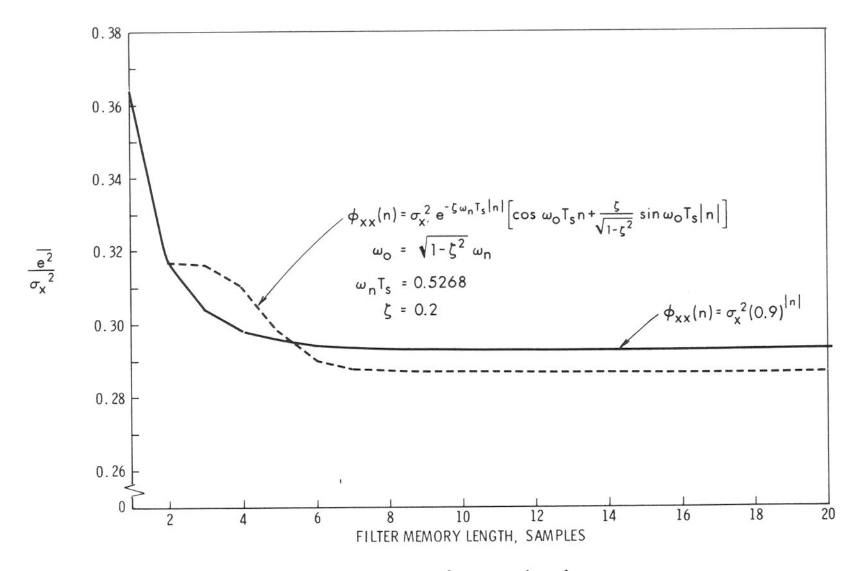

Figure 4.6 Mean square error for two-level measurements.

Numerical Results

The optimal linear filters were calculated for quantized measurements of the first-order and second-order processes described above. The performance for a memory length of 20 was used as an indication of the performance of the infinite-memory filter. In Figure 4.7 these results are compared with the ensemble performance of the Gaussian fit algorithm (in the pulse-code-modulation mode) for the first-order

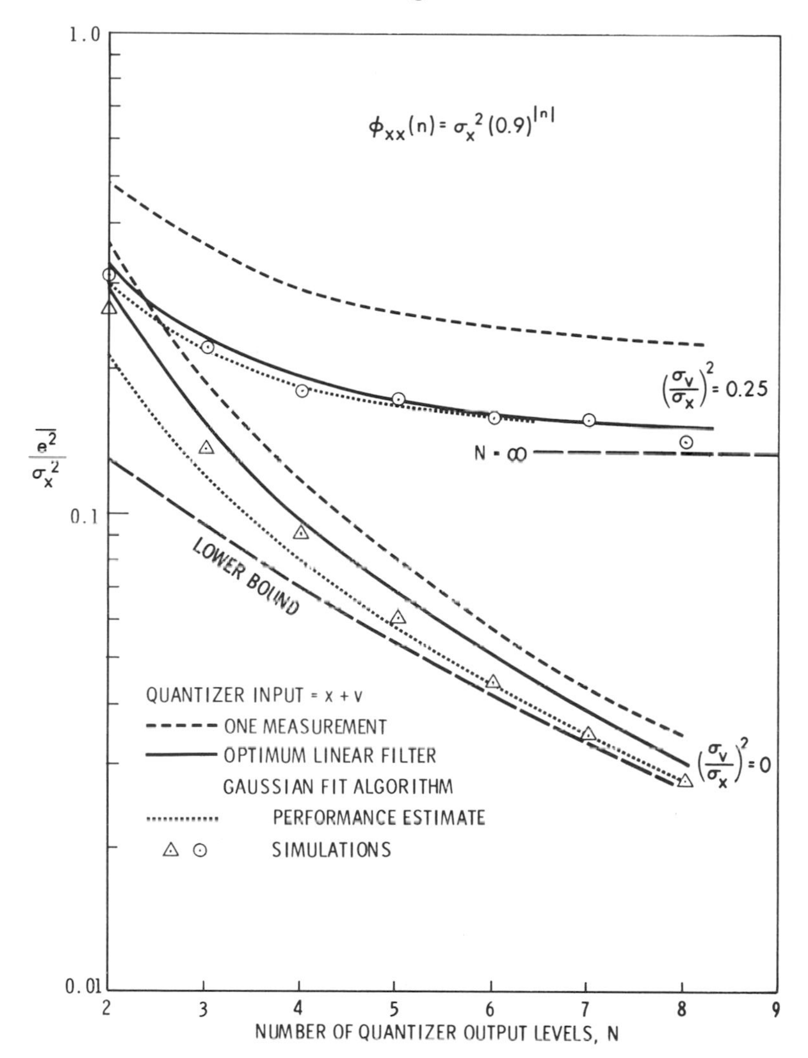

Figure 4.7 Estimation-error variance, Gaussian fit algorithm (PCM) and optimal linear filter.

process (the results with the second-order process were essentially the same and are not displayed). The graph shows data for cases both with and without observation noise.

In the case in which observation noise is included both filters show a rapid decrease in error variance with the number of quantum levels and at 8 output levels (3 bits) are within 12 percent of the error variance that is obtained with an infinite word length. There seems to be no advantage in performance over the linear filter beyond 4 or 5 output levels.

In the case in which observation noise is not included the error variance of the Gaussian fit algorithm is approximately 10 percent below that of the optimal linear filter for 2 to 8 output levels. When the two methods are compared for their ability to attain the lower bound, it is seen that for 4 or more output levels the linear filter remains roughly halfway between the upper bound (the one-measurement error variance) and the lower bound.* The Gaussian fit algorithm is rapidly approaching the lower bound and achieves more than 90 percent of the difference between upper and lower bounds by the time 8 output levels are used.

The problem of quantizers with linear feedback is discussed in Section 6.4, where it is shown that the optimal linear feedback element must be determined in an iterative manner. Although no performance estimates of quantizers with optimal linear feedback were made, it is expected that they too will compare favorably with the Gaussian fit algorithm when the input processes are stationary.

4.5 Joint Optimization of the Quantizer and Filter

Until this point the optimal Gaussian quantizer has been discussed. It is optimal in the sense that the quantum intervals and quantizer outputs have been so chosen as to minimize the mean square error between the quantizer input and output when the input sample is drawn from a Gaussian distribution.

There is no reason to think that this choice of parameters will be optimal when linear estimates of the quantizer input are being made from the quantizer output. Kellog (1967) briefly discusses the problem and points out that the optimal static (one sample) quantizer in conjunction with filtering is truly optimal only if the quantizer input samples are independent. Thus, we might expect that the optimal quantizer parameters will be a function of the filter memory length and the second-order probability-density functions of the quantizer input process.

We proceed by assuming that the optimal linear filter is used for any choice of quantizer parameters, and we find the necessary conditions for these parameters in terms of optimal filter weights. This is a complex

* This lower bound is computed under the assumptions that all measurements except the last one are linear and that only the last one is quantized.

parameter-optimization problem: the mean square error is a function of the quantizer describing-function gain k, the optimal linear filter weights, and the quantizer output autocorrelation function; all of these are functions of the quantizer parameters.

The usual approach in system design has been to use the optimal static quantizer (Kellog, 1967; Gish, 1967). In this section (and Appendix C) we give some analytical justification for this procedure: it is shown that ensemble mean square error is relatively insensitive to changes in quantizer parameters when the optimal static quantizer is used. This result is not restricted to Gaussian processes.

The Necessary Equations for Minimum Mean Square Error

In Section 4.4 the optimal linear filter $\boldsymbol{K}$ and mean square error in the optimal linear estimate were stated for zero-mean stationary processes:

$$\boldsymbol{K} = \mathsf{k}\boldsymbol{\phi}_{xx}^T \boldsymbol{Y}^{-1}, \tag{4.39}$$

$$\begin{aligned} \overline{e^2} &= \overline{x^2} - \boldsymbol{K}\boldsymbol{Y}\boldsymbol{K}^T \\ &= \overline{x^2} - \boldsymbol{\phi}_{xx}^T \left(\frac{\boldsymbol{Y}}{\mathsf{k}^2}\right)^{-1} \boldsymbol{\phi}_{xx}. \end{aligned} \tag{4.40}$$

where $\boldsymbol{\phi}_{xx}$ is an M-dimensional column vector whose ith component is $\phi_{xx}(i-1)$, $\boldsymbol{Y}$ is an $M \times M$ matrix whose ijth element is $\phi_{yy}(i-j)$, and M is the memory length of the linear filter.

The mean square error can be influenced by the quantizer parameters because it is a function of the quantizer output autocorrelation ϕ_{yy} and the quantizer describing-function gain k; these, in turn, are functions of the quantizer parameters and the second-order probability-density function of the quantizer input process. Define the $M \times M$ Toeplitz matrix Y as

$$\mathsf{Y} = \{\mathsf{Y}_{ij}\} = \left(\frac{\phi_{yy}(i-j)}{\mathsf{k}^2}\right). \tag{4.41}$$

This may be used in Equation 4.40 to give

$$\overline{e^2} = \overline{x^2} - \boldsymbol{\phi}_{xx}^T \mathsf{Y}^{-1} \boldsymbol{\phi}_{xx}. \tag{4.42}$$

Let $\delta\mathsf{Y}$ be the matrix of small changes in the elements of Y and $\delta\overline{e^2}$ the change in the mean square error that is caused by $\delta\mathsf{Y}$:

$$\begin{aligned} \delta\overline{e^2} &= -\boldsymbol{\phi}_{xx}^T \delta(\mathsf{Y}^{-1}) \boldsymbol{\phi}_{xx} \\ &= \boldsymbol{\phi}_{xx}^T \mathsf{Y}^{-1} \delta\mathsf{Y}\, \mathsf{Y}^{-1} \boldsymbol{\phi}_{xx}, \end{aligned} \tag{4.43}$$

But

$$\phi_{xx}^T \mathsf{Y}^{-1} = \mathsf{k}^2 \phi_{xx}^T Y^{-1} = \mathsf{k}\boldsymbol{K}. \tag{4.44}$$

Substituting Equation 4.44 in Equation 4.43 gives

$$\delta\overline{e^2} = \boldsymbol{K}\mathsf{k}^2 \delta \mathsf{Y} \boldsymbol{K}^T = \sum_{i=1}^{M} \sum_{j=1}^{M} K_i K_j \mathsf{k}^2 \delta \mathsf{Y}_{ij}. \tag{4.45}$$

There are only M independent elements in the $M \times M$ matrix Y, and they are the M quantities:

$$b(m) = \frac{\phi_{yy}(m)}{\mathsf{k}^2}, \qquad m = 0, 1, \ldots, M - 1. \tag{4.46}$$

It is shown in Appendix C that with the use of this information Equation 4.45 may be written

$$\delta\overline{e^2} = \left(\mathsf{k}^2 \sum_{i=1}^{M} K_i^2\right) \delta b(0) + \sum_{m=1}^{M-1} \left(2\mathsf{k}^2 \sum_{j=1}^{M-m} K_j K_{j+m}\right) \delta b(m). \tag{4.47}$$

As before, let the jth quantum interval A^j, with $j = 1, \ldots, N$, have lower and upper bounds d^j and d^{j+1}, respectively, and let y^j be the quantizer output when the quantizer input z falls in A^j.

Note that the mean square error is an explicit function only of the parameters $b(i)$, with $i = 0, \ldots, M - 1$, by virtue of Equations 4.41, 4.42, and 4.46, and the $b(i)$ in turn are functions of the quantizer parameters d^i and y^i. Thus the partial derivatives of the mean square error with respect to the quantizer parameters may be found by means of the chain rule:

$$\delta\overline{e^2} = \sum_{i=0}^{M-1} \frac{\partial\overline{e^2}}{\partial b(i)} \left(\sum_{n=1}^{N+1} \frac{\partial b(i)}{\partial d^n} \delta d^n + \sum_{n=1}^{N} \frac{\partial b(i)}{\partial y^n} \delta y^n \right)$$
$$= \sum_{n=1}^{N+1} \left[\sum_{i=0}^{M-1} \frac{\partial\overline{e^2}}{\partial b(i)} \left(\frac{\partial b(i)}{\partial d^n} \right) \right] \delta d^n + \sum_{n=1}^{N} \left[\sum_{i=0}^{M-1} \frac{\partial\overline{e^2}}{\partial b(i)} \left(\frac{\partial b(i)}{\partial y^n} \right) \right] \delta y^n. \tag{4.48}$$

From this the necessary equations for minimum mean square error are

$$
\begin{aligned}
&\sum_{i=0}^{M-1} \frac{\partial \overline{e^2}}{\partial b(i)}\left(\frac{\partial b(i)}{\partial d^n}\right) = 0, \qquad n = 1, 2, \ldots, N + 1, \\
&\sum_{i=0}^{M-1} \frac{\partial \overline{e^2}}{\partial b(i)}\left(\frac{\partial b(i)}{\partial y^n}\right) = 0, \qquad n = 1, 2, \ldots, N.
\end{aligned} \tag{4.49}
$$

The partial derivatives of the mean square error with respect to the variables $b(i)$ that appear here may be inferred from Equation 4.47:

$$
\frac{\partial \overline{e^2}}{\partial b(i)} = \begin{cases} \mathsf{k}^2 \sum_{j=1}^{M} K_j^2, & i = 0, \\ 2\mathsf{k}^2 \sum_{j=1}^{M-i} K_j K_{j+i}, & i = 1, 2, \ldots, M - 1. \end{cases} \tag{4.50}
$$

The partial derivatives of the $b(i)$ with respect to the quantizer parameters are derived in Appendix C and are

$$
\frac{\partial b(m)}{\partial d^n} = \begin{cases} \frac{2}{\mathsf{k}^2} p_z(d^n)(y^{n-1} - y^n)\left(\frac{y^{n-1} + y^n}{2} - \frac{\phi_{yy}(0)}{\mathsf{k}\sigma_z^2} d^n\right), & m = 0, \\ \frac{2}{\mathsf{k}^2} p_z(d^n)(y^{n-1} - y^n)\left[E(y_{j+m} \mid z_j = d^n) - \frac{\phi_{yy}(m)}{\sigma_z^2 \mathsf{k}} d^n\right], & m \neq 0, \end{cases} \tag{4.51}
$$

$$
\frac{\partial b(m)}{\partial y^n} = \frac{2}{\mathsf{k}^2} P(z \in A^n)\left[E(y_{j+m} \mid y_j = y^n) - \frac{\phi_{yy}(m)}{\sigma_z^2 \mathsf{k}} E(z_j \mid z_j \in A^n)\right)\Bigg], \tag{4.52}
$$

where z_j and y_{j+m} are the quantizer input and output at times t_j and t_{j+m}, respectively.

Substituting Equations 4.50, 4.51, and 4.52 in Equations 4.49 yields the specific necessary conditions for minimum mean square error in conjunction with optimal linear filtering. The number of equations is reduced when the probability distributions are symmetric.

Discussion

The necessary conditions that must be satisfied by the quantizer parameters in the one-sample, or static, case are given by Max (1960):

$$y^n = E(z \mid z \in A^n), \tag{4.53}$$

$$\frac{y^{n-1} + y^n}{2} = d^n. \tag{4.54}$$

These results may be substituted in Equations 4.51 and 4.52 to provide the partial derivatives of the $b(m)$ variables evaluated at the statically optimal quantizer parameters:

$$\frac{\partial b(m)}{\partial d^n} = \begin{cases} 0, & m = 0, \\ \frac{2}{\mathsf{k}^2} p_z(d^n)(y^{n-1} - y_n) \\ \quad \times \left[E(y_{j+m} \mid z_j = d^n) - \frac{\phi_{yy}(m)}{\overline{y^2}}\left(\frac{y^{n-1} + y^n}{2}\right)\right], & m \neq 0, \end{cases} \tag{4.55}$$

$$\frac{\partial b(m)}{\partial y^n} = \begin{cases} 0, & m = 0, \\ \frac{2}{\mathsf{k}^2} P(z \in A^n)\left(E(y_{j+m} \mid y_j = y^n) - \frac{\phi_{yy}(m)}{\overline{y^2}} y^n\right), & m \neq 0. \end{cases} \tag{4.56}$$

To obtain the last two equations the value of k that must be used is

$$\mathsf{k} = \sum_{i=1}^{N} \frac{y^i E(z \mid z \in A^i) P(z \in A^i)}{\sigma_z^2} = \sum_{i=1}^{N} \frac{(y^i)^2 P(z \in A^i)}{\sigma_z^2} = \frac{\overline{y^2}}{\sigma_z^2}. \tag{4.57}$$

It is expected that the partial derivatives of $b(m)$ with respect to the quantizer parameters will be small because of the following: (1) the bracketed term in Equation 4.55 is the difference between a conditional-mean estimate and the numerical average of two best linear estimates of the same quantity, and (2) the bracketed term in Equation 4.56 is the difference between a conditional-mean estimate and the best linear estimate of the same quantity.

This conjecture was verified numerically: not only did minimization involve extensive computation, but the mean square error was never reduced by more than 1 percent from the results obtained by using the statically optimal quantizer for the few cases that were tested.

We conclude that, since the statically optimal quantizer nearly satisfies the necessary conditions when filtering is included, it should be a good design in the majority of cases, although this cannot be guaranteed by the analysis presented here.

4.6 Summary

This chapter has dealt with the topic of optimal linear estimation with quantized measurements. The autocorrelation function of the quantizer output was examined in some detail for zero-mean Gaussian processes, and the quantizer was modeled by a gain element and an additive-noise source. Choosing the gain element as the random-input describing function not only minimizes the variance of the noise, but also minimizes the contribution of the noise autocorrelation to the output autocorrelation. This reduces the amount of computation required to obtain the output autocorrelation to any given accuracy. The describing function also provides the "whitest" noise source for two criteria of whiteness.

Computation of the optimal linear filter was described, and its ensemble performance was compared with that given by the Gaussian fit algorithm in the pulse-code-modulation mode. For the comparisons that were used it was found that the error variance of the Gaussian fit algorithm is roughly 10 percent lower than that of the optimal linear filter, although the Gaussian fit algorithm is much more efficient in attaining lower bounds. Lastly it was shown that if a quantizer is optimized without regard to the subsequent linear filter, it nearly satisfies the necessary conditions for the joint quantizer-filter optimization.

5 Optimal Stochastic Control with Quantized Measurements

5.1 Introduction

In this chapter we turn to the problem of controlling a system subject to random disturbances. Generally speaking, the problem of optimal control is more difficult to solve than that of optimal estimation. In fact, the latter can be shown to be a special case of the former. The optimal controller executes a strategy to minimize a performance index, which is the expected value of some function of the state variables and control variables. The optimal strategy combines the two functions of estimation and control, and the distinction between the two is not always clear. In some cases, usually involving the separation theorem, the tasks of the optimal stochastic controller can be divided into estimation and control.

The following section contains a statement of the problem of discrete-time optimal stochastic control. The solution through dynamic programming is outlined and applied to a two-stage process with a linear system and quadratic cost. The separation theorem is discussed. In the next section the optimal stochastic control is computed for a two-stage process with one quantized measurement, and the "dual control" theory of Fel'dbaum (1960) is used to explain the nonlinear control. Lastly, another separation theorem for nonlinear measurements is presented; the system is linear and the cost quadratic, and feedback around a nonlinear measurement device is required.

5.2 Optimal Stochastic Control

In this section we present a statement of the problem of optimal stochastic control for discrete-time systems and indicate the method of attack by dynamic programming. The solution for a two-stage process with a linear system, quadratic cost, and nonlinear measurements is obtained; the separation theorem is rederived by induction from this two-stage example.

Statement of the Problem

The problem of optimal stochastic control will now be stated. It is assumed that we know the following.

State equation:

$$\boldsymbol{x}_{i+1} = \boldsymbol{f}_i(\boldsymbol{x}_i, \boldsymbol{u}_i, \boldsymbol{w}_i),$$

Measurement equation:

$$\boldsymbol{z}_i = \boldsymbol{h}_i(\boldsymbol{x}_i, \boldsymbol{v}_i),$$

where

$\boldsymbol{x}_i$ = state vector at time t_i,
$\boldsymbol{u}_i$ = control vector at time t_i,
$\boldsymbol{z}_i$ = measurement vector at time t_i,
$\boldsymbol{w}_i$ = process-noise vector at time t_i,
$\boldsymbol{v}_i$ = measurement-noise vector at time t_i.

It is further assumed that the process-noise and observation-noise vectors are independent of each other and of the initial conditions, that their probability-density functions are known and, finally, that the probability-density function of the initial conditions is known. For clerical convenience let the collection of all past control actions and measurements be

$$\boldsymbol{\Theta}_k = \begin{cases} \boldsymbol{z}_k, \boldsymbol{z}_{k-1}, \ldots, \boldsymbol{z}_1, p(\boldsymbol{x}_0), \\ \boldsymbol{u}_{k-1}, \boldsymbol{u}_{k-2}, \ldots, \boldsymbol{u}_0. \end{cases} \tag{5.1}$$

For any given control policy $\boldsymbol{u}_i(\boldsymbol{\Theta}_i)$ the expected value of the cost function from time t_k to the end of the problem is given by

$$J_k = E\left\{ \sum_{j=k}^{N} L_j[\boldsymbol{x}_j, \boldsymbol{u}_j(\boldsymbol{\Theta}_j)] + \phi(\boldsymbol{x}_{N+1}) \middle| \boldsymbol{\Theta}_k \right\}. \tag{5.2}$$

As indicated in the equation, the expectation is conditioned on $\boldsymbol{\Theta}_k$ but, more importantly, it is carried out under the constraint that more measurements are to be taken in the future through the relationships $\boldsymbol{u}_j(\boldsymbol{\Theta}_j)$. The optimal control problem, then, is that of choosing the functional relations $\boldsymbol{u}_j(\boldsymbol{\Theta}_j)$ so as to minimize the expected cost of completing the process.

Solution by Dynamic Programming

The only known method of solution to this problem is dynamic programming (Dreyfus, 1965, 1964; Aoki, 1967; Bellman, 1962). As has been pointed out by several people (for example, Dreyfus, 1965), the cost is now a *functional* of the state probability-density function, as distinct from what it is in the deterministic-control problem, in which it is a *function* of the state.

Proceeding to the last stage, the remaining cost is

$$J_N = E[L_N(\boldsymbol{x}_N, \boldsymbol{u}_N) + \phi(\boldsymbol{x}_{N+1}) \mid \boldsymbol{\Theta}_N]. \tag{5.3}$$

Given any measurement and control sequence, this cost function is minimized with respect to $\boldsymbol{u}_N$, subject to the state equation constraint. The optimal cost at t_N, J_N^o, may then be expressed as

$$J_N^o = J_N^o(\boldsymbol{u}_0, \ldots, \boldsymbol{u}_{N-1}; z_1, \ldots, z_N) = J_N^o(\boldsymbol{\Theta}_N),$$

where it is to be understood that the arguments of J_N are those that determine the required conditional probability-density functions. At time t_{N-1} the cost to complete the process by using *any* $\boldsymbol{u}_{N-1}$, but then the *optimal* $\boldsymbol{u}_N^o$ is

$$J_{N-1} = E[L_{N-1}(\boldsymbol{x}_{N-1}, \boldsymbol{u}_{N-1}) + E(J_N^o \mid \boldsymbol{\Theta}_{N-1}, \boldsymbol{u}_{N-1}) \mid \boldsymbol{\Theta}_{N-1}]. \tag{5.4}$$

The expectation over all possible measurements z_N is necessary, since at time t_{N-1} z_N has not yet occurred. The control $\boldsymbol{u}_{N-1}$ is so chosen as to minimize J_{N-1}, and the process of stepping backward and minimizing is repeated. This leads to the well-known recursion relation (Dreyfus, 1965):

$$J_i^o(\boldsymbol{\Theta}_i) = \min_{\boldsymbol{u}_i} E[L_i(\boldsymbol{x}_i, \boldsymbol{u}_i) + E(J_{i+1}^o(\boldsymbol{\Theta}_{i+1}) \mid \boldsymbol{\Theta}_i, \boldsymbol{u}_i) \mid \boldsymbol{\Theta}_i]. \tag{5.5}$$

Linear Systems and Quadratic Cost

Now consideration will be limited to systems described by linear difference equations subject to a quadratic-cost criterion:

$$\boldsymbol{x}_{k+1} = \boldsymbol{\Phi}_k \boldsymbol{x}_k + \boldsymbol{G}_k \boldsymbol{u}_k + \boldsymbol{w}_k, \tag{5.6}$$

$$J_k = E\left[\sum_{i=k}^{N} \boldsymbol{x}_i^T \boldsymbol{A}_i \boldsymbol{x}_i + \boldsymbol{u}_i^T(\boldsymbol{\Theta}_i) \boldsymbol{B}_i \boldsymbol{u}_i(\boldsymbol{\Theta}_i) + \boldsymbol{x}_{N+1}^T \boldsymbol{S}_{N+1} \boldsymbol{x}_{N+1} \mid \boldsymbol{\Theta}_k\right]. \tag{5.7}$$

where the newly introduced quantities are the positive definite weighting matrices $\boldsymbol{A}_i$, $\boldsymbol{B}_i$, and $\boldsymbol{S}_{N+1}$, the state transition matrix $\boldsymbol{\Phi}_k$, and the control gain matrix $\boldsymbol{G}_k$. No constraints are placed on the measurement equation or control.

Let us now develop the solution to a two-stage problem, using the concepts outlined above. The total cost is

$$J_0 = E(\boldsymbol{x}_0^T \boldsymbol{A}_0 \boldsymbol{x}_0 + \boldsymbol{u}_0^T \boldsymbol{B}_0 \boldsymbol{u}_0 + \boldsymbol{x}_1^T \boldsymbol{A}_1 \boldsymbol{x}_1 + \boldsymbol{u}_1^T \boldsymbol{B}_1 \boldsymbol{u}_1 + \boldsymbol{x}_2^T \boldsymbol{S}_2 \boldsymbol{x}_2 \mid \boldsymbol{\Theta}_0). \tag{5.8}$$

The cost for the last stage is

$$\begin{aligned} J_1 &= E(\boldsymbol{x}_1^T \boldsymbol{A}_1 \boldsymbol{x}_1 + \boldsymbol{u}_1^T \boldsymbol{B}_1 \boldsymbol{u}_1 + \boldsymbol{x}_2^T \boldsymbol{S}_2 \boldsymbol{x}_2 \mid \boldsymbol{\Theta}_1) \\ &= E(\boldsymbol{x}_1^T \boldsymbol{A}_1 \boldsymbol{x}_1 + \boldsymbol{u}_1^T \boldsymbol{B}_1 \boldsymbol{u}_1 + \|\boldsymbol{\Phi}_1 \boldsymbol{x}_1 + \boldsymbol{G}_1 \boldsymbol{u}_1 + \boldsymbol{w}_1\|_{\boldsymbol{S}_2} \mid \boldsymbol{\Theta}_1). \end{aligned} \tag{5.9}$$

where the latter result is obtained by using the state equation, Equation 5.6, for $\boldsymbol{x}_2$. The following notation is used for the quadratic form:

$$\|\boldsymbol{y}\|_{(\;)} = \boldsymbol{y}^T(\ldots)\boldsymbol{y}.$$

By taking the derivative of the cost with respect to $\boldsymbol{u}_1$ or by completing squares it can be shown that the optimal choice of control $\boldsymbol{u}_1$ is $\boldsymbol{u}_1^o$:

$$\boldsymbol{u}_1^o = -(\boldsymbol{B}_1 + \boldsymbol{G}_1^T \boldsymbol{S}_2 \boldsymbol{G}_1)^{-1} \boldsymbol{G}_1^T \boldsymbol{S}_2 \boldsymbol{\Phi}_1 E(\boldsymbol{x}_1 \mid \boldsymbol{\Theta}_1). \tag{5.10}$$

Notice that this last control action is a linear function of the conditional mean of the state. This is always true for the last control action of a problem with linear system and quadratic cost; the two-stage problem considered here may be thought of as the last two stages of a control process of arbitrary length.

Equation 5.10 is substituted in Equation 5.9 to produce the optimal cost for the last stage:

$$\begin{aligned} J_1^o = {} & E[\boldsymbol{x}_1^T(\boldsymbol{A}_1 + \boldsymbol{\Phi}_1^T \boldsymbol{S}_2 \boldsymbol{\Phi}_1)\boldsymbol{x}_1 \mid \boldsymbol{\Theta}_1] + E\|\boldsymbol{w}_1\|_{\boldsymbol{S}_2} \\ & + E(\boldsymbol{x}_1^T \mid \boldsymbol{\Theta}_1)\boldsymbol{\Phi}_1^T \boldsymbol{S}_2 \boldsymbol{G}_1(\boldsymbol{B}_1 + \boldsymbol{G}_1^T \boldsymbol{S}_2 \boldsymbol{G}_1)^{-1} \boldsymbol{G}_1^T \boldsymbol{S}_2 \boldsymbol{\Phi}_1 E(\boldsymbol{x}_1 \mid \boldsymbol{\Theta}_1). \end{aligned} \tag{5.11}$$

We may drop the process-noise term from the cost function, since it cannot be influenced by control. Furthermore, we may define the following matrix for convenience:

$$\boldsymbol{C}_1 \equiv \boldsymbol{\Phi}_1^T \boldsymbol{S}_2 \boldsymbol{G}_1(\boldsymbol{B}_1 + \boldsymbol{G}_1^T \boldsymbol{S}_2 \boldsymbol{G}_1)^{-1} \boldsymbol{G}_1^T \boldsymbol{S}_2 \boldsymbol{\Phi}_1. \tag{5.12}$$

The last term of the cost function may be manipulated in the following manner:

$$\begin{aligned} E(\boldsymbol{x}_1^T \mid \boldsymbol{\Theta}_1)\boldsymbol{C}_1 E(\boldsymbol{x}_1 \mid \boldsymbol{\Theta}_1) &= \operatorname{tr}[\boldsymbol{C}_1 E(\boldsymbol{x}_1 \mid \boldsymbol{\Theta}_1) E(\boldsymbol{x}_1^T \mid \boldsymbol{\Theta}_1)] \\ &= \operatorname{tr}\{\boldsymbol{C}_1[E(\boldsymbol{x}_1\boldsymbol{x}_1^T \mid \boldsymbol{\Theta}_1) - E(\boldsymbol{e}_1\boldsymbol{e}_1^T \mid \boldsymbol{\Theta}_1)]\} \\ &= E(\boldsymbol{x}_1^T\boldsymbol{C}_1\boldsymbol{x}_1 \mid \boldsymbol{\Theta}_1) - \operatorname{tr}[\boldsymbol{C}_1 E(\boldsymbol{e}_1\boldsymbol{e}_1^T \mid \boldsymbol{\Theta}_1)], \end{aligned} \tag{5.13}$$

where tr(...) denotes the trace of a matrix, and $E(\boldsymbol{e}_1\boldsymbol{e}_1^T \mid \boldsymbol{\Theta}_1)$ is the covariance matrix of estimation errors *after* the measurement $\boldsymbol{z}_1$ has been processed. The optimal cost of completing the stage may be rewritten by substituting Equation 5.13 in Equation 5.11 (and dropping $E\|\boldsymbol{w}_1\|_{\boldsymbol{S}_2}$):

$$J_1^o = E(\boldsymbol{x}_1^T\boldsymbol{S}_1\boldsymbol{x}_1 \mid \boldsymbol{\Theta}_1) + \operatorname{tr}[\boldsymbol{C}_1 E(\boldsymbol{e}_1\boldsymbol{e}_1^T \mid \boldsymbol{\Theta}_1)], \tag{5.14}$$

where the matrix $\boldsymbol{S}_1$ is

$$\begin{aligned} \boldsymbol{S}_1 &= \boldsymbol{A}_1 + \boldsymbol{\Phi}_1^T\boldsymbol{S}_2\boldsymbol{\Phi}_1 - \boldsymbol{C}_1 \\ &= \boldsymbol{A}_1 + \boldsymbol{\Phi}_1^T[\boldsymbol{S}_2 - \boldsymbol{S}_2\boldsymbol{G}_1(\boldsymbol{B}_1 + \boldsymbol{G}_1^T\boldsymbol{S}_2\boldsymbol{G}_1)^{-1}\boldsymbol{G}_1^T\boldsymbol{S}_2]\boldsymbol{\Phi}_1. \end{aligned} \tag{5.15}$$

The cost for the last stage will now be averaged over all possible measurements $\boldsymbol{z}_1$:

$$\begin{aligned} E(J_1^o \mid \boldsymbol{\Theta}_0, \boldsymbol{u}_0) &= E\{E(\boldsymbol{x}_1^T\boldsymbol{S}_1\boldsymbol{x}_1 \mid \boldsymbol{\Theta}_1) + \operatorname{tr}[\boldsymbol{C}_1 E(\boldsymbol{e}_1\boldsymbol{e}_1^T \mid \boldsymbol{\Theta}_1)] \mid \boldsymbol{\Theta}_0, \boldsymbol{u}_0\} \\ &= E\{\boldsymbol{x}_1^T\boldsymbol{S}_1\boldsymbol{x}_1 + \operatorname{tr}[\boldsymbol{C}_1 E(\boldsymbol{e}_1\boldsymbol{e}_1^T \mid \boldsymbol{\Theta}_1)] \mid \boldsymbol{\Theta}_0, \boldsymbol{u}_0\}. \end{aligned} \tag{5.16}$$

Substituting this result in Equation 5.4 yields the total cost arrived at by using *any* initial control $\boldsymbol{u}_0$ and the *optimal* control $\boldsymbol{u}_1^o$:

$$\begin{aligned} J_0 = E\{&\boldsymbol{x}_0^T\boldsymbol{A}_0\boldsymbol{x}_0 + \boldsymbol{u}_0^T\boldsymbol{B}_0\boldsymbol{u}_0 + \boldsymbol{x}_1^T\boldsymbol{S}_1\boldsymbol{x}_1 \\ &+ \operatorname{tr}[\boldsymbol{C}_1 E(\boldsymbol{e}_1\boldsymbol{e}_1^T \mid \boldsymbol{\Theta}_1)] \mid \boldsymbol{\Theta}_0, \boldsymbol{u}_0\}. \end{aligned} \tag{5.17}$$

The optimal control $\boldsymbol{u}_0^o$ is that which minimizes this expression.

The Separation Theorem

At this point we shall discuss some sufficient conditions for the so-called separation theorem (Joseph and Tou, 1961; Deyst, 1967; Gunckel and Franklin, 1963). In addition to the previous assumptions about the independence of the process-noise vectors, observation-noise vectors, and initial-condition vector, it will now be assumed that they are normally distributed with known means and variances. It will also be assumed that the measurement equation is linear and that a measurement will be taken at each time instant.

Several pertinent points may now be made:

1. All random variables are normally distributed because of the linear equations and Gaussian initial conditions.
2. The conditional mean of the state vector is generated by the (linear) Kalman filter.
3. The covariance matrix of estimation errors conditioned on all past observations is known a priori (Deyst, 1967; Striebel, 1965) and is not influenced by the realization of the measurement/control sequence.

Dynamic programming must be used to solve for the optimal stochastic control sequence $\{u_0^o, u_1^o, \ldots, u_N^o\}$. The two-stage problem may be used to advantage by replacing t_2 with t_{N+1}, t_1 with t_N, and t_0 with t_{N-1}. Equations 5.9, 5.10, and 5.17 become

$$J_N = E(x_N^T A_N x_N + u_N^T B_N u_N + x_{N+1}^T S_{N+1} x_{N+1} \mid \Theta_N, u_N), \tag{5.18}$$

$$u_N^o = -(B_N + G_N^T S_{N+1} G_N)^{-1} G_N^T S_{N+1} \Phi_N \hat{x}_{N|N}, \tag{5.19}$$

$$\begin{aligned} J_{N-1} = {} & E(x_{N-1}^T A_{N-1} x_{N-1} + u_{N-1}^T B_{N-1} u_{N-1} \\ & + x_N^T S_N x_N \mid \Theta_{N-1}, u_{N-1}) + \operatorname{tr}[E(C_N E_{N|N} \mid \Theta_{N-1}, u_{N-1})], \end{aligned} \tag{5.20}$$

where

$$\begin{aligned} \hat{x}_{N|N} &= E(x_N \mid \Theta_N), \\ C_N &= \Phi_N^T S_{N+1} G_N (B_N + G_N^T S_{N+1} G_N)^{-1} G_N^T S_{N+1} \Phi_N, \\ S_N &= A_N + \Phi_N^T S_{N+1} \Phi_N - C_N, \\ E_{N|N} &= E[(x_N - \hat{x}_{N|N})(x_N - \hat{x}_{N|N})^T \mid \Theta_N]. \end{aligned}$$

Because the covariance matrix of estimation errors, $E_{N|N}$, is independent of any control action, the optimal control u_{N-1}^o may be determined from a cost function equivalent to Equation 5.20:

$$\begin{aligned} J'_{N-1} = {} & E(x_{N-1}^T A_{N-1} x_{N-1} + u_{N-1}^T B_{N-1} u_{N-1} \\ & + x_N^T S_N x_N \mid \Theta_{N-1}, u_{N-1}). \end{aligned} \tag{5.21}$$

Since this expression is the same as Equation 5.18 with its subscripts reduced by 1, it follows that the expression for optimal control, Equation 5.19, is valid if all its subscripts are reduced by 1. The same holds for the matrices C_N and S_N. The remaining optimal control actions are found in a like manner and may be expressed as

$$\boldsymbol{u}_i^o = -(\boldsymbol{B}_i + \boldsymbol{G}_i^T\boldsymbol{S}_{i+1}\boldsymbol{G}_i)^{-1}\boldsymbol{G}_i^T\boldsymbol{S}_{i+1}\boldsymbol{\Phi}_i\hat{\boldsymbol{x}}_{i|i}, \tag{5.22}$$

where

$$\boldsymbol{S}_i = \boldsymbol{A}_i + \boldsymbol{\Phi}_i^T[\boldsymbol{S}_{i+1} - \boldsymbol{S}_{i+1}\boldsymbol{G}_i(\boldsymbol{B}_i + \boldsymbol{G}_i^T\boldsymbol{S}_{i+1}\boldsymbol{G}_i)^{-1}\boldsymbol{G}_i^T\boldsymbol{S}_{i+1}]\boldsymbol{\Phi}_i,$$

$$\boldsymbol{S}_{N+1} \text{ given.} \tag{5.23}$$

The control gain matrix in Equation 5.22 is the same one that arises in the deterministic problem of a linear system and quadratic-cost control (Bryson and Ho, 1969). Thus the optimal stochastic control is "separated" into two parts: determination of the conditional mean of the state vector with the Kalman filter and multiplication of this estimate by the optimal gain sequence from the deterministic linear system and quadratic cost control problem.

5.3 Some Examples of Optimal Stochastic Control with Quantized Measurements

The purpose of this section is to compute the optimal stochastic control for a two-stage problem by using several quantizing elements in the observation path.

The scalar state under consideration obeys the following difference equation:

$$x_i = x_{i-1} + u_{i-1}, \qquad i = 1, 2. \tag{5.24}$$

The initial conditions are Gaussian with mean $\bar{x}_0$ and variance $\sigma_{x_0}^2$. No process noise has been included since, for the first stage, it may be combined with the initial conditions and, for the second (last) stage, it does not influence the control. One measurement without observation noise is to be taken at t_1 through a quantizer, which will be specified later. The cost function for this problem is

$$J_0 = E(b_0u_0^2 + a_1x_1^2 + b_1u_1^2 + a_2x_2^2 \mid \Theta_0), \tag{5.25}$$

where b_0, a_1, b_1, and a_2 are positive weighting coefficients corresponding to the matrices of Equation 5.8 (the mean square value of the initial state cannot be influenced by control and has been eliminated).

For linear systems and quadratic cost the optimal value of the last control action is always a linear function of the conditional mean of the state, as given by Equation 5.10:

$$u_1^o = -\frac{a_2}{b_1 + a_2}E(x_1 \mid \Theta_1). \tag{5.26}$$

Let c_1 and s_1 be the scalar versions of the matrices $\boldsymbol{C}_1$ (Equation 5.12) and $\boldsymbol{S}_1$ (Equation 5.15):

$$c_1 = \frac{a_2^2}{b_1 + a_2},$$
$$s_1 = a_1 + a_2 - \frac{a_2^2}{b_1 + a_2}. \tag{5.27}$$

The total cost when the optimal value of u_1^o is used may be written (see Equation 5.17) as

$$J_0 = E(b_0 u_0^2 + s_1 x_1^2 \mid \Theta_0, u_0) + c_1 E(E(e_1^2 \mid \Theta_1) \mid \Theta_0, u_0). \tag{5.28}$$

Let the M quantum intervals be denoted by A^i, with $i = 1, \ldots, M$. Substituting for x_1 from the state equation, Equation 5.24, the cost may be rewritten as

$$J_0 = s_1[\sigma_{x_0}^2 + (\bar{x}_0 + u_0)^2] + b_0 u_0^2 + c_1 \sum_{i=1}^{M} \operatorname{cov}(x_1 \mid x_1 \in A^i)\, P(x_1 \in A^i). \tag{5.29}$$

where $\operatorname{cov}(x_1 \mid x_1 \subset A^i)$ is the mean square estimation error in x_1, given that x_1 lies in A^i, and $P(x_1 \in A^i)$ is the a priori probability that x_1 will lie in A^i. Both these quantities are functions of the initial control, because x_1 is normally distributed with mean $\bar{x}_0 + u_0$ and variance $\sigma_{x_0}^2$.

It is convenient to define a quantizer distortion function D at this time:

$$D\left(\frac{\bar{z}}{\sigma_z}\right) = \frac{1}{\sigma_z^2} \sum_{i=1}^{M} \operatorname{cov}(z \mid z \in A^i)\, P(z \in A^i). \tag{5.30}$$

This is the normalized mean square estimation error when a random variable is quantized. If the limits of the quantum intervals are $\{d^i\}$,

$$z \in A^i, \qquad \text{if} \quad d^i \leq z < d^{i+1}$$

and the a priori distribution is Gaussian (mean $\bar{z}$, variance σ_z^2), then the distortion function may be explicitly stated by using the formulas in Chapter 2 for Gaussian random variables:

$$D\left(\frac{\bar{z}}{\sigma_z}\right) = \sum_{i=1}^{M} \left[P(z \in A^i) + \mu_i \frac{\exp(-\frac{1}{2}\mu_i^2)}{(2\pi)^{1/2}} - \mu_{i+1} \frac{\exp(-\frac{1}{2}\mu_{i+1}^2)}{(2\pi)^{1/2}} \right.$$

$$\left. - \frac{1}{P(z \in A^i)} \left(\frac{\exp(-\frac{1}{2}\mu_i^2)}{(2\pi)^{1/2}} - \frac{\exp(-\frac{1}{2}\mu_{i+1}^2)}{(2\pi)^{1/2}} \right)^2 \right], \tag{5.31}$$

$$P(z \in A^i) = \int_{\mu_i}^{\mu_{i+1}} \frac{\exp(-\frac{1}{2}v^2)}{(2\pi)^{1/2}}\, dv, \tag{5.32}$$

$$\mu_i = \frac{d^i - \bar{z}}{\sigma_z}.$$

The cost function, Equation 5.29, will be normalized with respect to $\sigma_{x_0}^2$, which yields

$$\frac{J_0}{\sigma_{x_0}^2} = s_1 \left[1 + \left(\frac{\bar{x}_0}{\sigma_{x_0}} + \frac{u_0}{\sigma_{x_0}} \right)^2 \right] + b_0 \left(\frac{u_0}{\sigma_{x_0}} \right)^2 + c_1 D \left(\frac{\bar{x}_0 + u_0}{\sigma_{x_0}} \right). \tag{5.33}$$

Although observation noise has not been included, it can be taken into account by conditioning the estimation error in Equation 5.29 on $z_1 \in A^i$, where

$$z_1 = x_1 + v_1,$$

$$E(v_1) = 0, \quad E(v_1^2) = \sigma_v^2. \tag{5.34}$$

The formulas for mean square estimation error conditioned on quantized Gaussian measurements were derived in Chapter 2. They may be used to show that with observation noise the normalized cost function becomes

$$\frac{J_0}{\sigma_{x_0}^2} = s_1 \left[1 + \left(\frac{\bar{x}_0}{\sigma_{x_0}} + \frac{u_0}{\sigma_{x_0}} \right)^2 \right] + b_0 \left(\frac{u_0}{\sigma_{x_0}} \right)^2$$

$$+ c_1 \frac{(\sigma_v/\sigma_{x_0})^2 + D[(\bar{x}_0 + u_0)/\sigma_z]}{1 + (\sigma_v/\sigma_{x_0})^2}, \tag{5.35}$$

where $\sigma_z^2 = \sigma_{x_0}^2 + \sigma_v^2$.

Observation noise decreases the dependence of the cost function on the distortion term in two ways. First, it is seen from Equation 5.35 that the distortion term D is multiplied by the quantity $\sigma_{x_0}^2/\sigma_z^2$, which is always less than 1 when observation noise is present. Second, the distortion function D should, generally speaking, be smaller for the larger input standard deviation σ_z, given the same quantum intervals defined by $\{d^i\}$, because the ratio of quantum interval size to standard deviation is smaller.

Numerical Results

The optimal control law was calculated by minimizing Equation 5.33 for three quantizers, shown in Figure 5.1; a two-level quantizer, a threshold quantizer, and a three-level quantizer. Observation noise was not included. The quantities u_0^2, x_1^2, and u_1^2 carried unit weighting for all computations. Two different weightings for the squared terminal error, x_2^2, were used: 1 and 100. The latter value presumes that the root-mean-square terminal error is ten times "more important" than the other three quantities.

In all cases in which the weighting of the terminal error was unity, the optimal stochastic control was extremely close (within graphical accuracy) to a straight line of slope -0.6. This line is the optimal control law for a linear measurement; it is not presented graphically. This somewhat surprising result may be explained by examining the cost function, Equation 5.33. The coefficients s_1, b_0, and c_1 have values

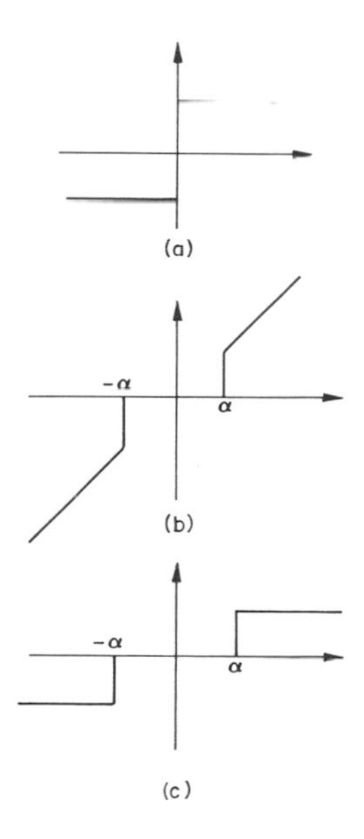

Figure 5.1 Quantizers used for observation. (a) Two-level quantizer; (b) threshold quantizer; (c) three-level quantizer.

1.5, 1.0, and 0.5, respectively. Since the distortion function D cannot exceed unity, that term never contributes more than 0.5 to the cost function and has very little influence on the choice of optimal control.

The optimal control laws for terminal-error weighting of 100 are shown in Figures 5.2, 5.3, and 5.4. The optimal (linear) control for linear measurements—that is, without quantization—is also shown. The deviation from the linear measurement case comes from the "dual control" aspect of optimal stochastic control (Fel'dbaum, 1960); the control application not only influences the state in the usual manner but also acts like a test signal to gain information about the system. For the specific problem at hand any error in the estimate of the state after the measurement has been processed at time t_1 will propagate through the system dynamics and contribute to the terminal error at t_2. Because the terminal error is heavily weighted, the control at time t_0 is used for setting up the measurement so that a more accurate estimate can be obtained. These facts are reflected in the cost function, Equation 5.33, where the coefficients s_1, b_0, and c_1 have values 2.0, 1.0, and 99.0, respectively.

The optimal control law for observations through a two-level quantizer is, for all practical purposes, linear with slope -0.95; see Figure 5.2. For the system used here a slope of -1.0 would center the distribution of x_1 at zero, which is the optimal situation for estimating

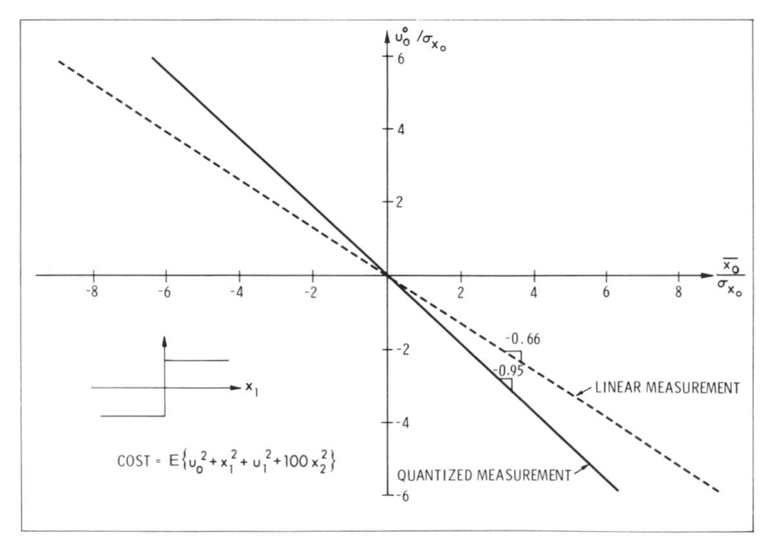

Figure 5.2 Optimal control law for two-level quantizer.

the quantizer input alone. Thus the control action makes a strong effort to minimize the covariance of estimation errors.

The optimal control law for the threshold quantizer is shown in Figure 5.3 for threshold halfwidths of 0.5, 1.0, 3.0, and 5.0. Because there is no observation noise, a measurement falling in the linear range yields an exact estimate of the state. The optimal control takes advantage of this fact by so placing the mean of the a priori distribution of x_1 as to increase the probability of a perfect measurement. The slope of the lines for threshold halfwidths α/σ_{x_0} of 1.0, 3.0, and 5.0 is close to -1.0, indicating that the mean of the a priori distribution of x_1 is situated near the same value, regardless of the initial condition $\bar{x}_0$.

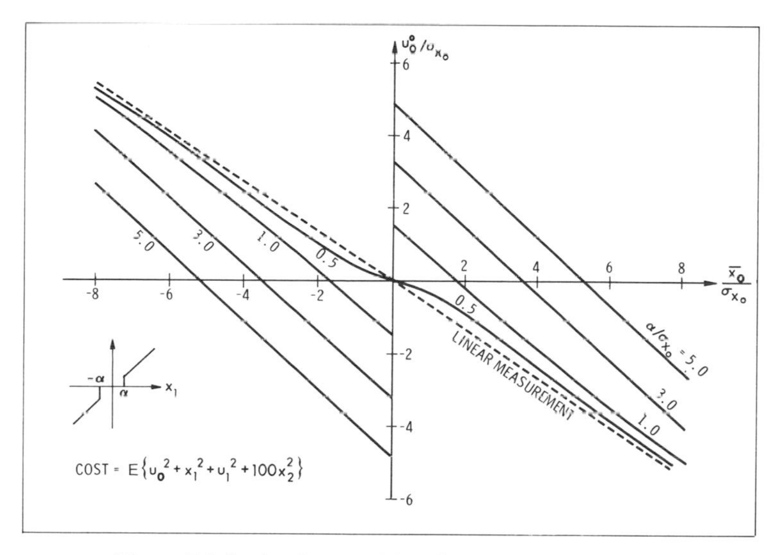

Figure 5.3 Optimal control law for threshold quantizer.

The curves for the three-level quantizer, Figure 5.4, resemble those for the threshold type of nonlinearity. They too have a slope near -1.0. For values of the threshold halfwidth of 0.5 and 1.0 the tendency is to put the distribution of x_1 at the center of the quantizer, to gain information. As the threshold halfwidth increases to $\alpha/\sigma_{x_0} = 3.5$, the distribution of x_1 is centered at a point close to $\pm\alpha$, the switching point, since each half of the three-level quantizer is effectively a two-level quantizer at such wide thresholds. The curve for $\alpha/\sigma_{x_0} = 5$ has the additional characteristic that it follows the linear-measurement control law for small values of $\bar{x}_0$. For larger values of the initial

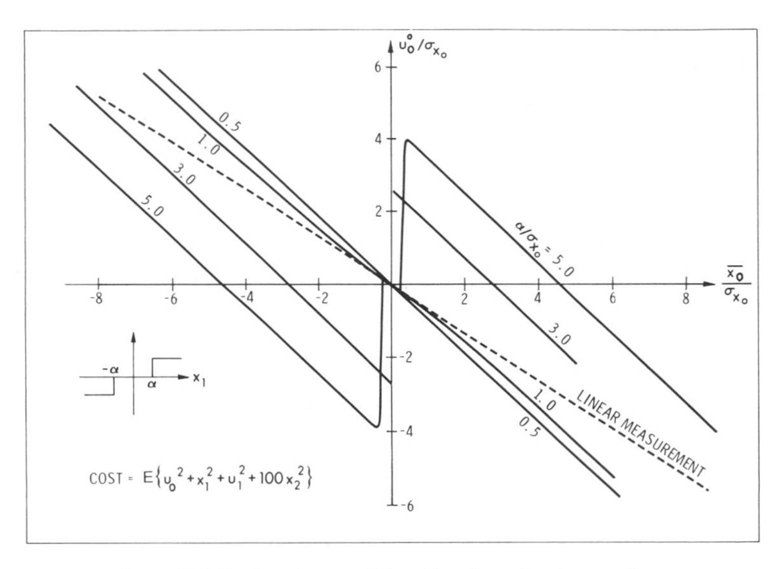

Figure 5.4 Optimal control law for three-level quantizer.

condition, however, the policy is changed, and the probability-density function of x_1 is centered near the switching point. It was found that with threshold halfwidths between 1.1 and 4.9 the policy is to go to the switching point if $\bar{x}_0$ is zero. With halfwidths larger than 4.9 the measurement opportunities are ignored, unless the initial condition is close enough to the switch point to make the control expenditure worth while.

The optimal control laws will not be as dramatic when the weighting is more evenly distributed. Nevertheless, the results presented here suggest methods of controlling a system when the measurements are quantized and the estimation errors are to be kept small. For example, a control action might be computed by ignoring the quantizer completely; this control could then be modified to place the mean value of the next measurement at or near the "closest" switching point of the quantizer or the midpoint of the quantum interval, whichever is preferable.

Processes with More than Two Stages

The difficulties inherent in the solution to the problem of optimal stochastic control with more than two stages can easily be described at this time. Recall that all the computations of the optimal control in the preceding example were done under the assumption that the distribution of the second from the last state variable (x_0) was normal. This is not true when quantized measurements have been taken prior to the determination of the next to the last control, and the expectations must be found by numerical integration, in general. To make the problem even more complicated, the distribution of the state contains all previous control actions as parameters; this implies, for example, that the first control action will influence nearly all terms of the cost function in a very complicated manner.

5.4 A Separation Theorem for Nonlinear Measurements

In this section we present the result that is displayed in Figure 5.5, which shows a linear system subject to quadratic cost with an arbitrary nonlinear element in the observation path. It will be demonstrated that if the proper feedback loop can be placed around the nonlinear element, the optimal stochastic controller is the cascading of (a) the nonlinear filter for generating the conditional mean of the state and (b) the optimal (linear) controller that is obtained when all noises are neglected.

Given:

$$\boldsymbol{x}_{i+1} = \boldsymbol{\Phi}_i \boldsymbol{x}_i + \boldsymbol{G}_i \boldsymbol{u}_i + \boldsymbol{w}_i, \qquad i = 1, 2, \ldots, N, \tag{5.36}$$

$$z_i = \boldsymbol{h}_i[(\cdot), \boldsymbol{v}_i], \tag{5.37}$$

$$J_k = E\left(\sum_{i=k}^{N+1} \boldsymbol{x}_i^T \boldsymbol{A}_i \boldsymbol{x}_i + \boldsymbol{u}_i^T \boldsymbol{B}_i \boldsymbol{u}_i \mid \boldsymbol{\Theta}_k \right), \tag{5.38}$$

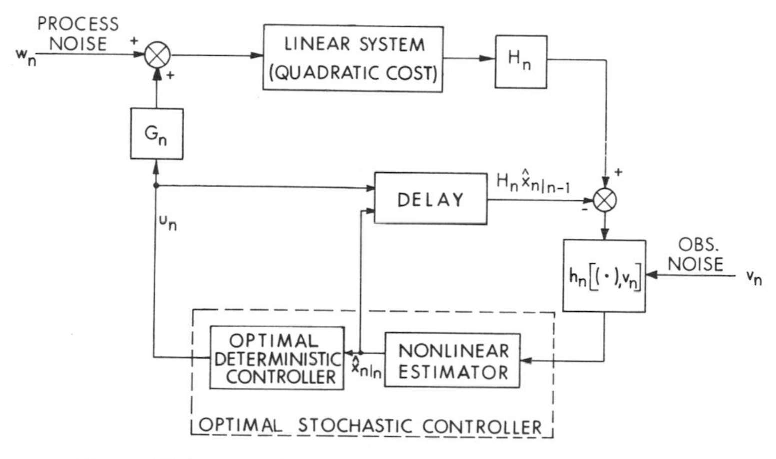

Figure 5.5 A separation theorem for nonlinear measurements.

$$\mathbf{\Theta}_k = \begin{cases} z_1, z_2, \ldots, z_k, p(\mathbf{x}_0), \\ \mathbf{u}_0, \mathbf{u}_1, \ldots, \mathbf{u}_{k-1}, \end{cases}$$

$$p(\mathbf{w}_i), \quad p(\mathbf{v}_i), \quad p(\mathbf{x}_0).$$

It is further assumed that the process-noise vectors, the measurement-noise vectors, and the initial-condition vector are independent of each other.

Now place a feedback loop around the nonlinear element, and subtract the mean of the incoming measurement conditioned on past data, thus forming the residual $\mathbf{r}_{k|k-1}$:

$$\begin{aligned} \mathbf{r}_{k|k-1} &= \mathbf{H}_k \mathbf{x}_k - E(\mathbf{H}_k \mathbf{x}_k \mid \mathbf{\Theta}_{k-1}, \mathbf{u}_{k-1}) \\ &= \mathbf{H}_k \mathbf{e}_{k|k-1}, \end{aligned} \tag{5.39}$$

where the prediction error $\mathbf{e}_{k|k-1}$ is given by

$$\begin{aligned} \mathbf{e}_{k|k-1} &= \mathbf{x}_k - E(\mathbf{x}_k \mid \mathbf{\Theta}_{k-1}, \mathbf{u}_{k-1}) \\ &= \mathbf{\Phi}_{k-1} \mathbf{e}_{k-1|k-1} + \mathbf{w}_{k-1}, \\ \mathbf{e}_{k-1|k-1} &= \mathbf{x}_{k-1} - E(\mathbf{x}_{k-1} \mid \mathbf{\Theta}_{k-1}). \end{aligned} \tag{5.40}$$

The measurement equation is modified so that now

$$z_k = \mathbf{h}_k(\mathbf{r}_{k|k-1}, \mathbf{v}_k). \tag{5.41}$$

When the conditional mean is subtracted from the incoming measurement, the control then has no influence upon the distribution of the quantity actually measured, and therefore it cannot affect the covariance of estimation errors. The detailed proof of the separation theorem is carried out in Appendix D, but by substituting Equations 5.39 and 5.40 in Equation 5.41 it can be seen that the distribution of z_k is determined by the distributions of $\mathbf{w}_{k-1}$, $\mathbf{v}_k$, and $\mathbf{e}_{k-1|k-1}$. Any function that does not include the control variables (but it may include past measurements) can be incorporated with the residual, Equation 5.39, and the separation character of the optimal-control computations is still valid.

The optimal control sequence, as calculated in Appendix D, is given by

$$\mathbf{u}_i^o = -(\mathbf{B}_i + \mathbf{G}_i^T \mathbf{S}_{i+1} \mathbf{G}_i)^{-1} \mathbf{G}_i^T \mathbf{S}_{i+1} \mathbf{\Phi}_i E(\mathbf{x}_i \mid \mathbf{\Theta}_i), \tag{5.42}$$

$$\mathbf{S}_i = \mathbf{A}_i + \mathbf{\Phi}_i^T [\mathbf{S}_{i+1} - \mathbf{S}_{i+1} \mathbf{G}_i (\mathbf{B}_i + \mathbf{G}_i^T \mathbf{S}_i \mathbf{G}_i)^{-1} \mathbf{G}_i^T \mathbf{S}_{i+1}] \mathbf{\Phi}_i. \tag{5.43}$$

$$\mathbf{S}_{N+1} = \mathbf{A}_{N+1}.$$

This control law is identical with Equation 5.22 in the case of linear measurements, in which the optimal stochastic controller is separated into two elements: a filter for generating the conditional mean of the state vector and a linear control law that is optimal when all uncertainties are neglected.

Computational Advantages

The primary advantage of feedback around the nonlinear measurement device is that the computation required to determine the optimal control sequence is greatly reduced: the backward solution (and associated storage) is eliminated, and the control action is a linear function of the conditional mean of the state. A nonlinear filter is required for finding the conditional mean. This may be a significant problem in its own right, but it will be easier to approximate a nonlinear filter than to obtain a solution by dynamic programming.

If the cost is not quadratic, dynamic programming must be used for finding the optimal control sequence (Deyst, 1967; Striebel, 1965). Feedback around the nonlinearity would simplify the computations here, too, since the probability-density functions of the measurements are independent of the control.

Performance

No claim has been made about the performance of the system that uses feedback around the nonlinear measurement device. Depending on the situation, the observation with feedback may give either better or worse performance than the optimal control without feedback. This may be explained by using the examples of two-level and threshold quantizers given in Section 5.3.

The feedback around the nonlinearity centers the probability-density function of the measurement at the origin of the quantizing element. For the threshold quantizer this means that the poorest measurement situation exists all the time, and it cannot be changed by control action. On the other hand, the optimal control action for measurements through a two-level quantizer tends to place the center of the probability-density function of the measurement at the switching point. But feedback around the quantizer centers the distribution "free of charge," that is, without expenditure of control, and the total cost is less than that when feedback is not used.

The situation with feedback around the threshold quantizer may be improved by introducing a bias at the quantizer input to ensure an accurate measurement. The separation theorem introduced in this section is still valid, and the cost can be reduced. Indeed, one can extend this concept to a whole new class of problems in stochastic control, in which a set of measurement control variables may be considered in addition to the state equation control variable. For quantized measurements these variables could be a bias or a scale factor or both, much as in the case of the quantizer with feedback, Section 3.2. Meier and his associates (1967, b) consider controls that influence measurements, but they concentrate on linear measurements with controllable observation-noise statistics.

Applications

The measurement feedback could be implemented in any control system in which the linear system output exists in amplitude-continuous form before it is passed through a quantizer or other nonlinearity. The use of a nonlinear device would be dictated by hardware constraints or other considerations, since a linear (and optimal) control system could be designed for the available linear measurements. One example of a possible application is in inertial navigation, in which the accelerometer output exists in analog form before it is quantized and stored; feedback around the quantizer could be utilized in such a case (Quagliata, 1968).

5.5 Summary

In this chapter we considered the problem of optimal stochastic control and emphasized the case of linear systems, quadratic performance indices, and nonlinear measurements. The sufficient conditions for the separation of the estimation and control functions were described. A two-stage example was presented, and the nonlinear character of the optimal stochastic control is due to its "probing" property. The difficulties in treating processes of more than two stages were discussed. Lastly, it was shown that feedback around an arbitrary nonlinear measurement device can result in sufficient conditions for a separation theorem: the optimal stochastic controller is implemented by cascading, first, a nonlinear estimator, to generate the conditional mean of the state and, second, the optimal (linear) controller when all uncertainties are neglected. Perhaps the most immediate application is in systems in which the measurements undergo analog-to-digital conversion.

6 Suboptimal and Linear Control with Quantized Measurements

6.1 Introduction

In most cases of practical interest it is exceedingly difficult to obtain the solution to the optimal stochastic control problem (or the combined estimation and control problem) when the plant or measurement equations, or both, are nonlinear, when the controls are constrained, or when the cost is nonquadratic. This situation was demonstrated in Chapter 5, in which the results of only a two-stage problem with one quantized measurement were given. Different suboptimal control algorithms have evolved as a compromise between computational effort and an efficient control scheme; this chapter describes some previous work and some new ideas in this direction.

One widely used algorithm is the so-called open-loop-optimal feedback control (Dreyfus, 1964). In Section 6.2 this is applied to systems with nonlinear measurements, and it is compared with the truly optimal control described in the previous chapter. A new algorithm for suboptimal stochastic control is presented in Section 6.3. This algorithm has the "dual control" characteristics and can be applied to a wide range of problems. The design of optimal linear controllers for stationary closed-loop systems is treated in Section 6.4. A plant control system and the predictive-quantization system are discussed.

6.2 Open-Loop-Optimal Feedback Control

Dreyfus (1964) gives a particularly lucid description of the algorithm for open-loop-optimal feedback control. The operation is as follows. Measurements and control actions are recorded up to, say, time t_k, and an estimate of the state, $\hat{\boldsymbol{x}}_{k|k}$, is generated. It is temporarily assumed that no future measurements are to be taken, and a control sequence, $\{\boldsymbol{u}_k, \boldsymbol{u}_{k+1}, \ldots, \boldsymbol{u}_N\}$, is computed, to minimize the cost of completing the process. This is the optimal open-loop control sequence from time t_k, given past observations and control actions. Only the first member of the sequence, $\boldsymbol{u}_k$, is actually implemented, however; this advances the time index to $k + 1$, when a new measurement is taken, a new estimate $\hat{\boldsymbol{x}}_{k+1|k+1}$ is computed, and the process of computing the optimal open-loop control is repeated.

This open-loop-optimal feedback control law has the primary advantage that, although not optimal for the original problem, it is readily computed. Moreover, feedback is incorporated, so that it takes advantage of current information about the system.

Let us turn to the problem of controlling a linear system with a quadratic-cost criterion:

$$\begin{aligned}
\boldsymbol{x}_{k+1} &= \boldsymbol{\Phi}_k \boldsymbol{x}_k + \boldsymbol{G}_k \boldsymbol{u}_k + \boldsymbol{w}_k, \\
J_k &= E\left(\sum_{i=k}^{N+1} \boldsymbol{x}_i^T \boldsymbol{A}_i \boldsymbol{x}_i + \boldsymbol{u}_i^T \boldsymbol{B}_i \boldsymbol{u}_i \right), \\
E(\boldsymbol{w}_i \boldsymbol{w}_j^T) &= \boldsymbol{Q}_i \delta_{ij}.
\end{aligned} \tag{6.1}$$

To simplify manipulations somewhat, let

$$\begin{aligned}
\boldsymbol{Z}_k &= \{z_k, z_{k-1}, \ldots, z_1; p(\boldsymbol{x}_0)\}, \\
\boldsymbol{U}_{k-1} &= \{\boldsymbol{u}_{k-1}, \boldsymbol{u}_{k-2}, \ldots, \boldsymbol{u}_0\}, \\
\hat{\boldsymbol{x}}_{j|k} &= E(\boldsymbol{x}_j \mid \boldsymbol{Z}_k, \boldsymbol{U}_{j-1}).
\end{aligned}$$

Note that

$$\hat{\boldsymbol{x}}_{j+1|k} = \boldsymbol{\Phi}_j \hat{\boldsymbol{x}}_{j|k} + \boldsymbol{G}_j \boldsymbol{u}_j, \qquad j \geq k, \tag{6.2}$$

$$\boldsymbol{x}_{j+1} - \hat{\boldsymbol{x}}_{j+1|k} = \boldsymbol{e}_{j+1|k} = \boldsymbol{\Phi}_j \boldsymbol{e}_{j|k} + \boldsymbol{w}_j, \tag{6.3}$$

$$E(\boldsymbol{e}_{j|k} \mid \boldsymbol{Z}_k, \boldsymbol{U}_{j-1}) = 0.$$

Let

$$E(\boldsymbol{e}_{j|k} \boldsymbol{e}_{j|k}^T \mid \boldsymbol{Z}_k, \boldsymbol{U}_{j-1}) = \boldsymbol{E}_{j|k}$$

so that

$$E_{j+1|k} = \Phi_j E_{j|k} \Phi_j^T + Q_j. \tag{6.4}$$

The open-loop cost function, $J_{OL,k}$, the cost of completing the process from time t_k without incorporating future measurements, is given by

$$J_{OL,k} = E\left(\sum_{i=k}^{N+1} x_i^T A_i x_i + u_i B_i u_i \,\middle|\, Z_k, U_{k-1} \right),$$

$$J_{OL,k} = \sum_{i-k}^{N+1} E[(\hat{x}_{i|k} + e_{i|k})^T A_i (\hat{x}_{i|k} + e_{i|k})] + u_i^T B_i u_i$$

$$= \sum_{i=k}^{N+1} [\hat{x}_{i|k}^T A_i \hat{x}_{i|k} + u_i^T B_i u_i + \mathrm{tr}(A_i E_{i|k})]. \tag{6.5}$$

The open-loop-optimal feedback control law chooses the sequence $\{u_k, \ldots, u_N\}$ so as to minimize the cost function (Equation 6.5) subject to constraints of state (Equation 6.2) and covariance (Equation 6.4). For a comparison recall that the truly optimal stochastic control is found by choosing the functional relations $u_i(\Theta_i) = u_i(Z_i, U_{i-1})$ to minimize the expected cost of completing the process. Here we are, in effect, choosing $u_i(Z_k, U_{i-1})$, with $i \geq k$, to minimize the cost.

Since future measurements are being ignored, the covariance terms in the cost are completely determined by $E_{k|k}$ through Equation 6.4 and therefore cannot be controlled. Thus, an equivalent problem statement is as follows. Choose $u_k, \ldots, u_N$ to minimize $J'_{OL,k}$:

$$J'_{OL,k} = \sum_{i=k}^{N+1} \hat{x}_{i|k}^T A_i \hat{x}_{i|k} + u_i^T B_i u_i \tag{6.6}$$

subject to

$$\hat{x}_{i+1|k} = \Phi_i \hat{x}_{i|k} + G_i u_i, \qquad i \geq k, \tag{6.7}$$

where $\hat{x}_{k|k}$ is given. This is a deterministic control problem, whose solution is well known (see, for example, Bryson and Ho, 1969):

$$u_i^o = -(B_i + G_i^T S_{i+1} G_i)^{-1} G_i^T S_{i+1} \Phi_i \hat{x}_{i|k}, \tag{6.8}$$

$$S_i = A_i + \Phi_i^T [S_{i+1} - S_{i+1} G_i (B_i + G_i^T S_{i+1} G_i)^{-1} G_i^T S_{i+1}] \Phi_i,$$

$$S_{N+1} = A_{N+1}. \tag{6.9}$$

As mentioned previously, only the first member of the sequence is implemented; thus, for any time t_k the open-loop-optimal feedback control algorithm chooses a control $\boldsymbol{u}_k$ such that

$$\boldsymbol{u}_k = -(\boldsymbol{B}_k + \boldsymbol{G}_k^T \boldsymbol{S}_{k+1} \boldsymbol{G}_k)^{-1} \boldsymbol{G}_k^T \boldsymbol{S}_{k+1} \boldsymbol{\Phi}_k \hat{\boldsymbol{x}}_{k|k}. \tag{6.10}$$

This is the precise form of the control when the separation theorem is valid (Section 5.2). In other words, the open-loop-optimal feedback control proceeds as though the conditions for the separation theorem were valid, even though they are not.

When the open-loop-optimal feedback control sequence is substituted in the equivalent cost function, Equation 6.6, it becomes (Bryson and Ho, 1969):

$$(J'_{OL,k})^o = \hat{\boldsymbol{x}}_{k|k}^T \boldsymbol{S}_k \hat{\boldsymbol{x}}_{k|k}. \tag{6.11}$$

This in turn may be used in the open-loop cost function, Equation 6.5, to yield

$$J_{OL,k}^o = \hat{\boldsymbol{x}}_{k|k}^T \boldsymbol{S}_k \hat{\boldsymbol{x}}_{k|k} + \sum_{i=k}^{N+1} \operatorname{tr}(\boldsymbol{A}_i \boldsymbol{E}_{i|k}). \tag{6.12}$$

Open-Loop-Optimal Feedback Control with Quantized Measurements

In this section we shall apply the open-loop-optimal feedback control law to the quantized-measurement problems considered in Section 5.3. For that scalar state problem the open-loop-optimal feedback control, Equation 6.10, is

$$u_0 = -\frac{s_1}{s_1 + b_0} \bar{x}_0, \tag{6.13}$$

where $s_1 = a_1 + a_2 - a_2^2/(b_1 + a_2)$; see Section 5.3. This control law is the same as that used for linear measurements. It gives very good results when all the weighting coefficients in the cost function, Equation 5.25, are unity: the fractional increase in minimal cost,

$$\frac{J_{OL}^o - J^o}{J^o}$$

is always less than 0.05 percent for the examples presented in Section 5.3.

When the terminal-error weighting coefficient is increased to 100, the results are quite different, as is to be expected. Figures 6.1 and 6.2 show the fractional increase in cost when the open-loop-optimal feedback control is used for the threshold quantizer and for the two-level and three-level quantizers, respectively. The increase in cost is roughly 20 to 80 percent but may be as high as 170 percent for the threshold quantizer.

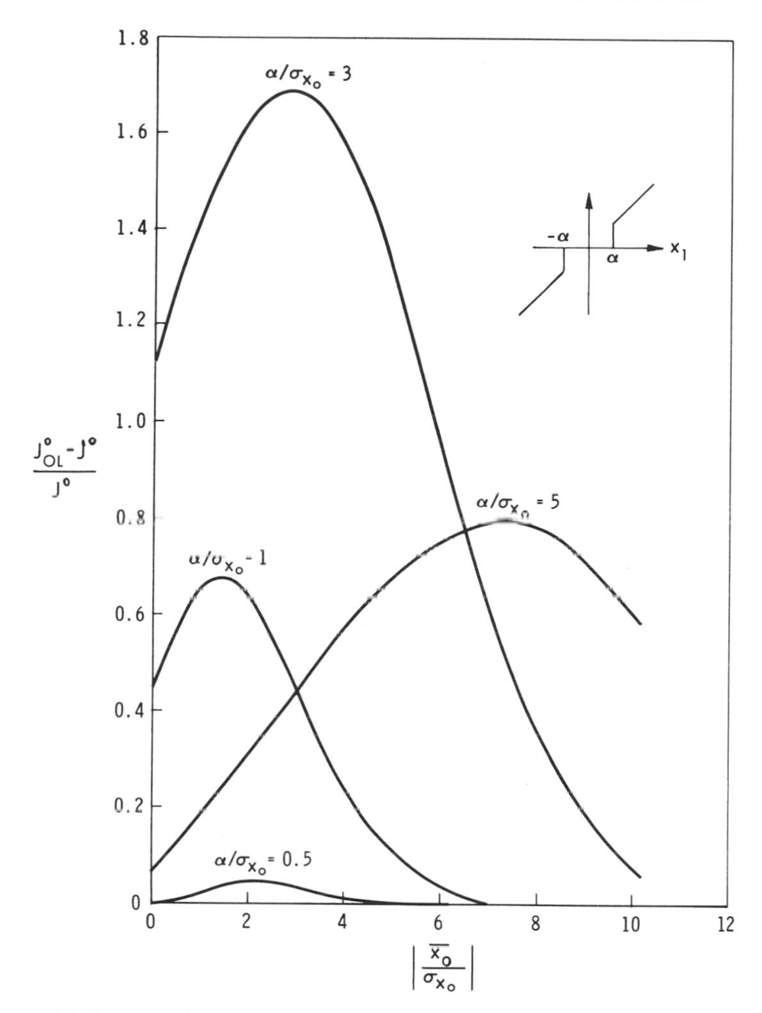

Figure 6.1 Fractional increase in minimal cost with open-loop-optimal feedback control.

The reason for this is evident by now: because it does not incorporate future measurements, the open-loop-optimal feedback control neglects the distortion term D in Equation 5.33 and the conditional-covariance term in Equation 6.5. It is precisely this term which is important when the terminal error is heavily weighted. In other words, the initial control is not expended in an effort to improve the quality of the measurement of x_1. This is especially serious when x_2 is heavily

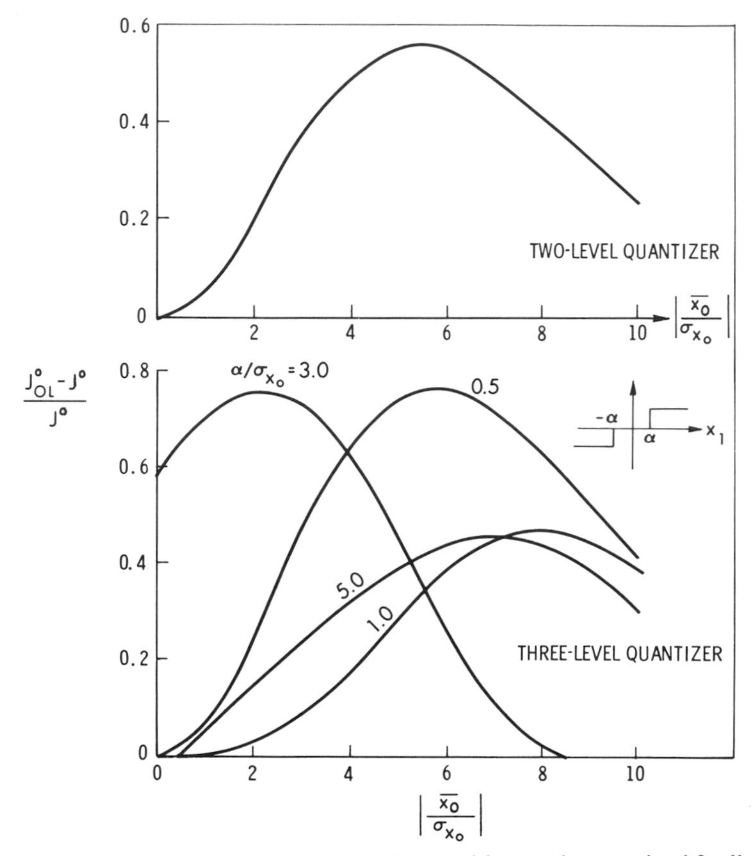

Figure 6.2 Fractional increase in minimal cost with open-loop-optimal feedback control: two-level and three-level quantizers.

weighted in the cost function, because any error in the estimate *after* the measurement has been processed propagates through the system and contributes to the terminal error x_2.

6.3 A New Algorithm for Suboptimal Stochastic Control

The open-loop-optimal feedback control law suffers the disadvantage that it does not account for future measurements. As shown in Section 6.2 for linear systems and quadratic cost, the open-loop cost function may be decomposed into two parts: a part involving the conditional mean and a part involving the conditional covariance. It is the latter part that is neglected by the open-loop-optimal feedback control.

The logical step toward improving the performance (and increasing the computational requirements) is to allow for one or more, say M, future measurements in the computation of the control law and to use the open-loop-optimal feedback control at all other times. We shall call this scheme M-measurement-optimal feedback control. First the algorithm is described and discussed, and the special case of one future measurement is examined in more detail. The section concludes with an example and some Monte Carlo simulation results of the stochastic control of a linear plant with coarsely quantized (nonlinear) measurements.

Description of the Control Algorithm

The algorithm proceeds as follows. Measurements and control actions are recorded up to, say, time t_k, and an estimate of the state vector (and, perhaps, its conditional probability-density function) is made. It is temporarily assumed that M measurements are to be taken in the future (at preassigned times). The control sequence $\{\boldsymbol{u}_k, \boldsymbol{u}_{k+1}, \ldots\}$ is so computed as to minimize the cost of completing the process taking M measurements along the way. This control sequence is the M-measurement-optimal control sequence. Only $\boldsymbol{u}_k$, the first member of the sequence, is actually implemented; this advances the time index, a new measurement is taken, a new estimate is computed, and the process of computing a new M-measurement-optimal control sequence is repeated.

The M-measurement-optimal feedback control has two limiting forms: when M is zero, no future measurements are included in the control computations, and it becomes the open-loop-optimal feedback control, but when M is the actual number of measurements remaining in the problem, it becomes the truly optimal stochastic control.

The primary advantage of the M-measurement-optimal feedback control over other suboptimal control algorithms is that one or more future observations are incorporated into the control computations. This allows for the "probing" or "test signal" type of control action, which is characteristic of the truly optimal stochastic control. The open-loop-optimal feedback control cannot take this form, since future measurements are ignored.

The concept of incorporating M future measurements rather than the actual number of measurements may also be used to simplify computations involving constrained controls, nonlinear plants, and nonquadratic cost criteria. For example, if the actual controls are constrained at all future times, it can be assumed for the purposes of control computation that they are constrained only at M future times and unconstrained at all other times. Similarly, it can be assumed that the plant is nonlinear only at M future times and linear at all other times, or that the cost is nonquadratic at M future times and quadratic at all other times. In all these situations the suboptimal control approaches the optimal control as M approaches the number of stages remaining in the problem. Generally speaking, the computational effort will increase as M increases; however, the writer knows of no analysis that shows that the performance will improve as M increases.

Problem Statement

Given the following.

(a) The state equation:

$$\boldsymbol{x}_{i+1} = \boldsymbol{f}_i(\boldsymbol{x}_i, \boldsymbol{u}_i, \boldsymbol{w}_i), \qquad i = 0, 1, \ldots, N, \tag{6.14}$$

where $\boldsymbol{x}_i$ is the state vector at time t_i, $\boldsymbol{u}_i$ is the control vector at time t_i, and $\boldsymbol{w}_i$ is the process noise at time t_i.

(b) The measurement equation:

$$\boldsymbol{z}_i = \boldsymbol{h}_i(\boldsymbol{x}_i, \boldsymbol{v}_i), \qquad i = 1, 2, \ldots, N, \tag{6.15}$$

where $\boldsymbol{v}_i$ is the observation noise at time t_i.

(c) The cost criterion:

$$J = E\left[\sum_{i=0}^{N} L_i(\boldsymbol{x}_i, \boldsymbol{u}_i) + \phi(\boldsymbol{x}_{N+1})\right]. \tag{6.16}$$

(d) The probability-density functions:

$$p(\boldsymbol{x}_0), \quad \{p(\boldsymbol{w}_i)\}, \quad \{p(\boldsymbol{v}_i)\}.$$

It is assumed that the vectors $\boldsymbol{x}_0$, $\{\boldsymbol{w}_i\}$, and $\{\boldsymbol{v}_i\}$ are independent.

Open-Loop-Optimal Feedback Control

The cost function to be minimized by the open-loop control at time t_k is given by

$$J_{OL,k} = E\left[\sum_{j=k}^{N} L_j(\boldsymbol{x}_j, \boldsymbol{u}_j) + \phi(\boldsymbol{x}_{N+1}) \,\middle|\, \boldsymbol{Z}_k, \boldsymbol{U}_{k-1}\right]. \tag{6.17}$$

This is minimized by the control sequence $\{\boldsymbol{u}_k, \ldots, \boldsymbol{u}_N\}$ subject to the state constraint, Equation 6.14. The first member of this sequence, $\boldsymbol{u}_k$, is the open-loop-optimal feedback control at time t_k.

One-Measurement-Optimal Feedback Control

We shall treat a special case of M-measurement-optimal feedback control. If only one future measurement is incorporated in the control computation at time t_k, and if this measurement occurs at time t_{k+n}, then the one-measurement cost function becomes

$$J_{OM,k} = E\left[\sum_{j=k}^{k+n-1} L_j(\boldsymbol{x}_j, \boldsymbol{u}_j) + E(J^o_{OL,k+n} \mid \boldsymbol{Z}_k, \boldsymbol{U}_{k+n-1}) \mid \boldsymbol{Z}_k, \boldsymbol{U}_{k-1}\right]. \tag{6.18}$$

In this equation the open-loop-optimal cost function $J^o_{OL,k+n}$ is similar to Equation 6.17 after minimization, but it depends on the measurements $\{\boldsymbol{Z}_k, z_{k+n}\}$ and the controls $\boldsymbol{U}_{k+n-1}$. The first member of the control sequence $\{\boldsymbol{u}_k, \ldots, \boldsymbol{u}_N\}$ that minimizes Equation 6.18 is the one-measurement-optimal feedback control at time t_k.

The solution may still be difficult to find in practice. Even when the explicit dependence of the open-loop cost function on the open-loop-optimal controls $\boldsymbol{u}_{k+n}, \ldots, \boldsymbol{u}_N$ can be removed, the solution of Equation 6.18 still involves an n-stage two-point boundary value problem. If, however, the future measurement occurs at the "next" time, then it reduces to a parameter optimization over $\boldsymbol{u}_k$; that is, the one-measurement cost function becomes

$$J_{OM,k} = E[L_k(\boldsymbol{x}_k, \boldsymbol{u}_k) + E(J^o_{OL,k+1} \mid \boldsymbol{Z}_k, \boldsymbol{U}_k) \mid \boldsymbol{Z}_k, \boldsymbol{U}_{k-1}]. \tag{6.19}$$

In this equation $J^o_{OL,k+1}$ depends on the measurements $\boldsymbol{Z}_{k+1}$ and the controls $\boldsymbol{U}_k$.

Linear Systems, Quadratic Cost, and Nonlinear Measurements

The following paragraphs contain some specialized results when the state equation is linear,

$$\boldsymbol{x}_{i+1} = \boldsymbol{\Phi}_i \boldsymbol{x}_i + \boldsymbol{G}_i \boldsymbol{u}_i + \boldsymbol{w}_i, \qquad i = 0, 1, 2, \ldots, N \tag{6.20}$$

and when the cost is quadratic,

$$J = E\left(\sum_{i=0}^{N} \boldsymbol{x}_i^T \boldsymbol{A}_i \boldsymbol{x}_i + \boldsymbol{u}_i^T \boldsymbol{B}_i \boldsymbol{u}_i + \boldsymbol{x}_{N+1}^T \boldsymbol{A}_{N+1} \boldsymbol{x}_{N+1}\right). \tag{6.21}$$

Here the weighting matrices $\{\boldsymbol{A}_i\}$ and $\{\boldsymbol{B}_i\}$ are positive semidefinite and positive definite, respectively. The measurements may still be nonlinear.

It is shown in Appendix E that the optimal open-loop cost function in this case may be rewritten

$$J^o_{OL,k+1} = E(\boldsymbol{x}^T_{k+1}\boldsymbol{S}_{k+1}\boldsymbol{x}_{k+1} \mid \boldsymbol{Z}_{k+1}, \boldsymbol{U}_k) + \text{tr}[\boldsymbol{E}_{k+1|k+1}(\boldsymbol{F}_{k+1} - \boldsymbol{S}_{k+1})] + \text{const}, \tag{6.22}$$

where

$$\boldsymbol{S}_i = \boldsymbol{A}_i + \boldsymbol{\Phi}^T_i[\boldsymbol{S}_{i+1} - \boldsymbol{S}_{i+1}\boldsymbol{G}_i(\boldsymbol{B}_i + \boldsymbol{G}^T_i\boldsymbol{S}_{i+1}\boldsymbol{G}_i)^{-1}\boldsymbol{G}^T_i\boldsymbol{S}_{i+1}]\boldsymbol{\Phi}_i,$$

$$\boldsymbol{S}_{N+1} = \boldsymbol{A}_{N+1}, \tag{6.23}$$

$$\boldsymbol{F}_i = \boldsymbol{\Phi}^T_i\boldsymbol{F}_{i+1}\boldsymbol{\Phi}_i + \boldsymbol{A}_i,$$

$$\boldsymbol{F}_{N+1} = \boldsymbol{A}_{N+1}, \tag{6.24}$$

$$\boldsymbol{E}_{k+1|k+1} = \text{cov}(\boldsymbol{x}_{k+1} \mid \boldsymbol{Z}_{k+1}, \boldsymbol{U}_k). \tag{6.25}$$

The equation for the one-measurement cost function with the future measurements occurring at the next time, Equation 6.19, becomes

$$J_{OM,k} = E[\boldsymbol{x}^T_k\boldsymbol{A}_k\boldsymbol{x}_k + \boldsymbol{u}^T_k\boldsymbol{B}_k\boldsymbol{u}_k + \boldsymbol{x}^T_{k+1}\boldsymbol{S}_{k+1}\boldsymbol{x}_{k+1} + E\{\text{tr}[\boldsymbol{E}_{k+1|k+1}(\boldsymbol{F}_{k+1} - \boldsymbol{S}_{k+1})] \mid \boldsymbol{Z}_k, \boldsymbol{U}_k\} \mid \boldsymbol{Z}_k, \boldsymbol{U}_{k-1}]. \tag{6.26}$$

The value of $\boldsymbol{u}_k$ that minimizes this is the one-measurement-optimal feedback control at time t_k for linear systems, quadratic cost, and nonlinear measurements. It is shown in Appendix E that the weighting matrix $\boldsymbol{F}_{k+1} - \boldsymbol{S}_{k+1}$ is at least positive semidefinite. If it were not, the control would minimize the cost by degrading, not improving, the knowledge of the state.

It still may be a difficult problem to compute the conditional expectations in Equation 6.26. The simulation results presented in Chapter 3 indicate, however, that in the case of quantized measurements, a good approximation is given by the Gaussian fit algorithm, which assumes that the prediction error at time t_{k+1} based on all past data will be Gaussian. It follows that the distribution of z_{k+1} before the measurement is taken also will be Gaussian, so the Gaussian formulas may be used for computing the expectation $\boldsymbol{E}_{k+1|k+1}$. In fact, using the Gaussian fit algorithm with the one-measurement-optimal feedback control is analogous to finding the first control action in a two-stage process with Gaussian initial conditions (cf. Equations 6.26 and 5.17).

A comment about steady-state operation: the nonlinear control law obtained by minimizing Equation 6.26 with respect to $\boldsymbol{u}_k$ will not, in general, reach a steady state even though the matrices $\boldsymbol{S}_{k+1}$, $\boldsymbol{F}_{k+1}$, $\boldsymbol{A}_k$, and $\boldsymbol{B}_k$ are constant. As discussed in Chapter 2, this results from the fact that the covariance matrix conditioned on quantized measurements, $\boldsymbol{E}_{k+1|k+1}$, cannot be expected to reach a steady-state value.

Simulations

Consider the scalar state equation given by

$$x_{k+1} = x_k + u_k + w_k, \qquad k = 0, 1, \ldots, N, \tag{6.27}$$

where x is the state, u is the scalar control, and w is Gaussian process noise. The initial conditions also have a Gaussian distribution. The cost function is quadratic in nature:

$$J = E\left(\sum_{i=0}^{N} A_i x_i^2 + B_i u_i^2 + A_{N+1} x_{N+1}^2\right). \tag{6.28}$$

Observations are taken through a three-level quantizer, shown in Figure 5.1, with switch points at $\pm\alpha$.

Obtaining numerical solutions to the problem of stochastic control requires extensive use of the probability-density function of the state conditioned on past measurements and control actions. Here it is assumed that the distribution just prior to a quantized measurement is Gaussian, as discussed above.

Both the open-loop-optimal feedback control and the one-measurement-optimal feedback control have been simulated on the digital computer. The latter algorithm assumes that the one future measurement occurs at the next time instant. The control algorithms used the same realizations of the initial conditions and process noise, although the state trajectories, in general, were different. Each case was run fifty times, and the ensemble averages were approximated by the numerical average of these fifty trials. In every case the weighting coefficients for the control effort $\{B_i\}$ are unity, and the process-noise variance is constant, but the latter varies from case to case.

Numerical Results

The results of a seven-stage stochastic process are described. Quantitative results are displayed for a terminal-control problem, and the results of other simulations are discussed.

Terminal Control In the terminal-control problem all state weighting coefficients are unity except the last one, A_7, which is 100. This means, roughly, that the root-mean-square terminal error is ten times "more important" than the other quantities in the cost function.

Figure 6.3 shows the results when the measurements are taken through a three-level quantizer; the ensemble mean square state and ensemble average of the (approximate) conditional covariance of the state are plotted as a function of time. In this case the variance of the process noise is 0.2, and the quantizer switch points are at ± 1.

The most noticeable difference between the two control laws is that the one-measurement control acts to reduce the conditional covariance of the state estimate. Note that the ensemble average of the conditional covariance is about half the average of the conditional covariance for the open-loop control. The one-measurement control is able to do this by centering the conditional distribution of the measurement near the quantizer switch point. This is reflected in the curves for the mean square value of the state, which stays in the neighborhood of 1.0 (the switch point) for the one-measurement control but gradually goes to zero for the open-loop control. The control effort (not shown) for the one-measurement control is higher, and it requires a large

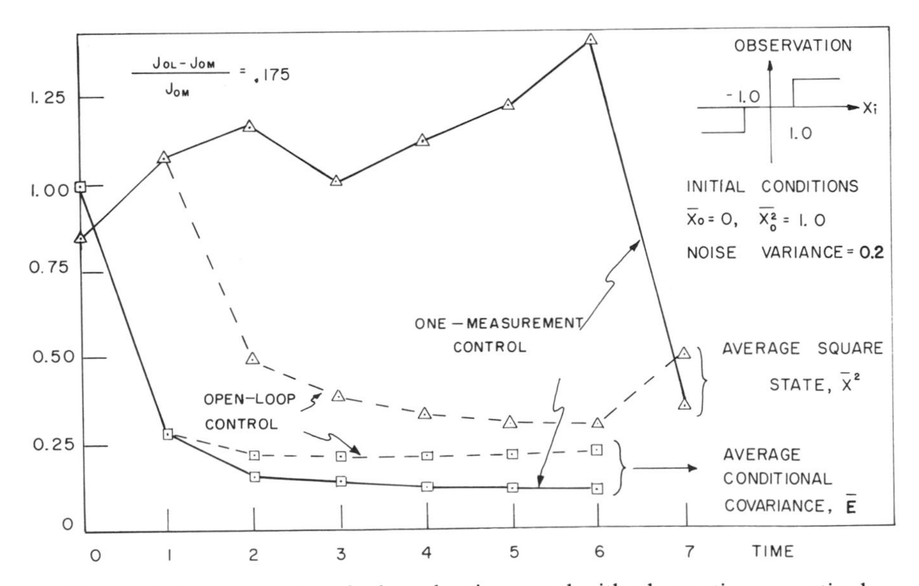

Figure 6.3 Seven-stage terminal stochastic control with observations quantized to three levels.

control action at the last application to bring the state from the vicinity of the quantizer switch point to the origin.

The performance penalty of the open-loop-optimal feedback control as against the one-measurement-optimal feedback control is 17 percent in this case. Other simulations revealed that the performance penalty ranged as high as 44 percent when observations were taken through a two-level quantizer.

Other Simulations Cost functions other than that of the terminal-control type were simulated: the state deviations were weighted more heavily as time progressed, or else the weightings were constant. The performance advantage of the one-measurement control was always less than 10 percent in these cases. This arises from the fact that the one-measurement control tries to move the state around to gain information, but these movements are restricted by the heavy weighting on the state deviations.

Thus, a qualitative assessment, at least for linear systems and non-linear measurements, is that incorporating future measurements in the control computations will yield the greatest return when the cost function is such that the state or control or both are free to reduce uncertainty in the estimate. In other situations the open-loop control is quite attractive, especially because of its computational simplicity.

6.4 The Best Linear Controller: Discussion

This section briefly discusses the problems inherent in the design of closed-loop systems when the measurements are quantized. A plant-control problem and a transmitter–receiver from a communication system (predictive quantization) are both reduced to an iterative solution of the Wiener filtering problem when the quantizer is replaced with a gain plus additive, uncorrelated noise.

Throughout this section the lowercase letter z denotes the Z-transform variable.

Plant Control

Figure 6.4a shows the block diagram of the control system in which the output is to follow the input random process x. Here e is the system error, Q is the quantizing element, y is the quantizer output, $D(z)$ is the digital compensation to be chosen, w is additive process noise, $F(z)$ is the fixed, time-invariant, stable pulse transfer function of the plant (and any hold networks), and c is the plant output. Synchronous

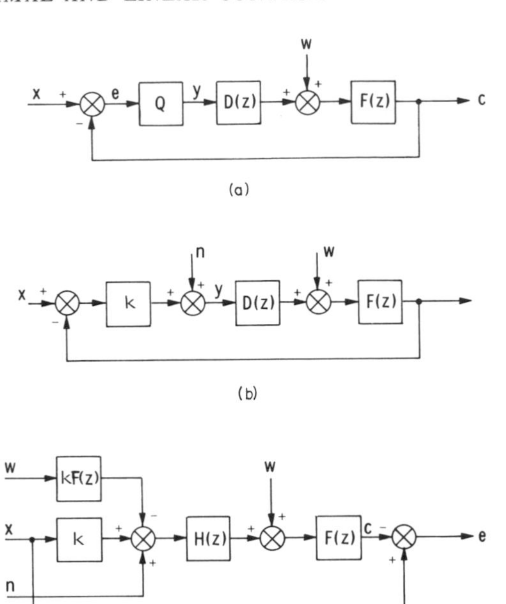

Figure 6.4 Plant-control block diagrams.

sampling elements are assumed to exist between all transfer functions.

We shall use the results of Chapter 4 and replace the quantizer with the describing-function gain k and an additive, uncorrelated noise source n. The quantizer output y then becomes

$$y = \mathrm{k}e + n, \tag{6.29}$$

where

$$\mathrm{k} = \frac{E(ey)}{E(e^2)}. \tag{6.30}$$

This results in an equivalent block diagram of Figure 6.4b. The equation for the Z transform of the output sequence c can then be written as

$$c(z) = \frac{D(z)F(z)}{1 + \mathrm{k}D(z)F(z)}\mathrm{k}x(z) + \frac{D(z)F(z)}{1 + \mathrm{k}D(z)F(z)}n(z)$$
$$+ \frac{F(z)}{1 + \mathrm{k}D(z)F(z)}w(z). \tag{6.31}$$

Define the transfer function $H(z)$ as

$$H(z) = \frac{D(z)}{1 + \mathrm{k}D(z)F(z)}. \tag{6.32}$$

Substituting this in Equation 6.31 yields

$$c(z) = F(z)H(z)[\mathrm{k}x(z) + n(z)] + \frac{F(z)H(z)}{D(z)}w(z). \tag{6.33}$$

$D(z)$ may be eliminated from the last term by using Equation 6.32:

$$c(z) = F(z)H(z)[\mathrm{k}x(z) + n(z)] + [1 - \mathrm{k}F(z)H(z)]F(z)w(z) \tag{6.34}$$

or, finally,

$$c(z) = F(z)H(z)[\mathrm{k}x(z) + n(z) - F(z)\mathrm{k}w(z)] + F(z)w(z). \tag{6.35}$$

The last equation suggests the block diagram of Figure 6.4c, which is a Wiener filtering problem of the semifree configuration type, (Vander Velde, 1967). The design problem is to find $H(z)$, to minimize the mean square value of the system error e. An iterative design process has to be used, because the optimal filter $H(z)$ depends on the spectrum of the noise n, which depends on the quantizer input e, which in turn depends on the filter $H(z)$. This is illuminated by the expression for the system error as follows. The system output, Equation 6.35, is subtracted from x:

$$e(z) = [1 - \mathrm{k}F(z)H(z)][x(z) - F(z)w(z)] - F(z)H(z)n(z). \tag{6.36}$$

In this equation $n(z)$ depends on $e(z)$.

The second-order probability distribution of the quantizer input must be known before the quantization-noise autocorrelation function can be calculated. In view of the filtering that takes place around the loop it is usually justifiable to assume that the error samples e are normally distributed. Smith (1966) indicates that this will be a good approximation if the spectrum of the input signal has a bandwidth wider than the linear part of the system. The addition of Gaussian process noise (not considered by Smith, 1966) will also make the quantizer input more Gaussian. It will be assumed from here on that the quantizer input samples are jointly normal.

The describing-function coefficient k will remain constant during the design process if the ratio of the quantum interval to the standard deviation is constant. This means that the quantum intervals and quantizer output values are not specified until the design process has been completed and the standard deviation of the quantizer input is known.

The Design Procedure The design procedure is purely heuristic. Discussions with H. Smith indicate that the results should be interpreted with some care. It may be initiated by computing the value of k for the quantizer being used. Quantization noise is neglected, and a function $H(z)$ that minimizes the mean square error $\overline{e^2}$ is chosen.

The autocorrelation function of the error e is computed* and, under the Gaussian hypothesis, this determines the autocorrelation of the quantization noise (Chapter 4). A new function $H(z)$ that includes quantization noise is chosen, and the procedure outlined in this paragraph is repeated. When a satisfactory $H(z)$ has been computed, the compensation $D(z)$ is then found from Equation 6.32. An alternative approach is to deal directly with the quantizer output autocorrelation ϕ_{yy} rather than the quantization-noise autocorrelation.

It should be noted that the optimal filter $H(z)$ will not be a rational function of z even if the input process x and noise process w have rational spectra, unless the quantization-noise spectrum is approximated by a shaping filter.

Judged by the successful experience with optimal linear filters in Chapter 4, it may well be that matrix methods rather than spectral factorization is an efficient way to solve this iterative design problem.

A Transmitter–Receiver System: Predictive Quantization

Figure 6.5a shows a transmitter–receiver pair that has been the subject of much investigation (for example: Bello et al., 1967; Davisson, 1966, 1967; Fine, 1964; Gish, 1967; O'Neal, 1966, 1968). It is desired that information about the input process $x_i = x(t_i)$ be sent over a communication channel. Linear feedback is placed around the quantizer, so that the signal being quantized is, in some sense, the error in predicting the incoming sample x_i. The quantizer input is u_i and the output y_i. The receiver operates on the sequence $\{y_i\}$ to produce an estimate of the transmitter input x_i.

When the quantization element is a relay and $D(z)$ is a pulse counter, the transmitter is the well-known delta modulation scheme. With

* Although the quantization noise is uncorrelated with the quantizer input, it is correlated with the x and w processes. The cross-power spectral density function between the "input" $x(z) - F(z)w(z)$ and the quantization noise $n(z)$ is required and may be found in terms of the quantization-noise spectral density by multiplying both sides of Equation 6.36 by $n(z)$, taking expectations, and setting the left-hand side equal to zero.

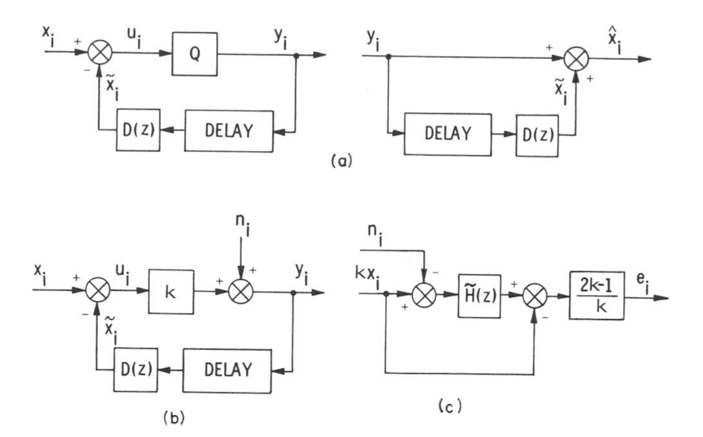

Figure 6.5 Transmitter-receiver block diagrams.

quantizers of more than two levels the system is sometimes called "differential pulse code modulation" (O'Neal, 1966). Using a threshold type of quantizer is mathematically equivalent to using a predictive-correction data-compression scheme (Davisson, 1967).

Both Fine (1964) and Gish (1967) have shown that if the receiver operation is to be linear, the transmitter feedback must be the same linear operation, in order to minimize the mean square estimation error. To determine this error we note in Figure 6.5a that

$$x_i = \tilde{x}_i + u_i, \tag{6.37}$$

$$\hat{x}_i = \tilde{x}_i + y_i, \tag{6.38}$$

where $\tilde{x}_i$ is the prediction of x_i based on $\{y_{i-1}, y_{i-2}, \ldots\}$. Subtracting the second of these equations from the first shows the error e_i to be

$$e_i = x_i - \hat{x}_i = -(y_i - u_i) \tag{6.39}$$

which is just the error incurred in quantizing u_i. Again we replace the quantizer with the describing-function gain k and an uncorrelated noise source n:

$$y_i = \mathrm{k}u_i + n_i, \tag{6.40}$$

where

$$\mathrm{k} = \frac{E(uy)}{E(u^2)}. \tag{6.41}$$

The error, Equation 6.39, may be written

$$e_i = (1 - \mathsf{k})u_i - n_i. \tag{6.42}$$

An expression for the Z transform of the quantizer input may be written by examining the equivalent block diagram for the transmitter, Figure 6.5b, as

$$u(z) = \frac{1}{1 + \mathsf{k}D(z)z^{-1}}x(z) + \frac{D(z)z^{-1}}{1 + \mathsf{k}D(z)z^{-1}}n(z). \tag{6.43}$$

Define the transfer function $H(z)$ as

$$H(z) = \frac{1}{1 + \mathsf{k}D(z)z^{-1}}. \tag{6.44}$$

The Z transform of u, Equation 6.43, may then be expressed as

$$u(z) = H(z)x(z) + \frac{1}{\mathsf{k}}[1 - H(z)]n(z). \tag{6.45}$$

Substituting this in Equation 6.42 gives the Z transform of the system error:

$$e(z) = -\frac{2\mathsf{k} - 1}{\mathsf{k}}\left(n(z) - \frac{\mathsf{k} - 1}{2\mathsf{k} - 1}H(z)[n(z) - \mathsf{k}x(z)]\right). \tag{6.46}$$

It is seen that, when displayed in this form, the filter $(\mathsf{k} - 1)H(z)/(2\mathsf{k} - 1)$ is trying to extract the "signal" $n(z)$ from an input of signal plus "noise," $-\mathsf{k}x(z)$. This result, perhaps surprising, is not inconsistent. The quantization-noise spectrum will have a wider bandwidth than will the spectrum of x. The optimal choice of $H(z)$ will therefore attenuate the low frequencies. This is realized by integration (summation) in the feedback compensation $D(z)$, an approach that was used by De Jaeger (see Gish, 1967). He analyzed and tested the system with sinusoidal inputs.

Another interpretation can be made by adding and subtracting $x(z)$ to the bracketed term in Equation 6.46. The equation for the system error then becomes

$$e(z) = -\frac{2\mathsf{k} - 1}{\mathsf{k}}\{\mathsf{k}x(z) - \tilde{H}(z)[\mathsf{k}x(z) - n(z)]\}, \tag{6.47}$$

where

$$\tilde{H}(z) = 1 - \frac{k - 1}{2k - 1} H(z). \tag{6.48}$$

This suggests the Wiener filtering problem shown in Figure 6.5c.

The design procedure for this situation is iterative, as in the plant-control problem, because the spectrum of quantization noise $n(z)$ depends upon the quantizer input, Equation 6.43. All assumptions and comments about the design process for the plant-control problem apply equally well to the quantizer with feedback considered here.

6.5 Summary

In this chapter we have investigated suboptimal control algorithms and linear controllers. The open-loop-optimal feedback control was compared with the truly optimal stochastic control in the two-stage example in Chapter 5. The penalty in performance associated with the open-loop control can be quite high and is due to the fact that this control neglects future measurements. A new algorithm, called M-measurement-optimal feedback control, was introduced to overcome the deficiencies of the open-loop control. It has the "probing" characteristic of the truly optimal control and can be applied in situations with other than nonlinear measurements. Simulations of the open-loop and M measurement controls indicate that the latter will yield the greatest improvement in performance when the cost function does not penalize the probing action too heavily, as in a terminal-control problem. The last topic dealt with closed-loop systems containing quantizers. Both a plant-control problem and the predictive-quantization system were restated as Wiener filtering problems requiring an iterative solution.

7 Summary and Recommendations for Further Study

7.1 Introduction

The mathematical operation of quantization exists in many communication and control systems. The increasing demand on existing digital facilities (such as communication channels and data-storage) can be alleviated by representing the same amount of information with fewer bits at the expense of more sophisticated data-processing. The purpose of this monograph is to examine the two distinct but related problems of state variable estimation and control when the measurements are quantized.

7.2 Summary and Conclusions

Analytical expressions have been derived for the conditional mean and conditional covariance of the state of a Gaussian linear system when the measurements have been arbitrarily quantized. These equations do not depend on the repeated application of Bayes' rule to the state probability-density function. The characteristics of the optimal filtering, smoothing, and prediction solutions were retained in the approximate implementations. The Gaussian fit filtering and smoothing algorithms appear to be the most promising because of their accuracy and ability to work with arbitrary quantization schemes and nonstationary processes. These techniques were applied to three digital communication systems: pulse-code-modulation, predictive quantization, and predictive-comparison data-compression. Simulation results compare quite favorably with performance estimates.

The important problem of designing linear estimators for quantized measurements of stationary processes was investigated. The quantizer is replaced with a gain element and an additive-noise source; the random-input describing function optimizes several criteria that simplify system analysis and synthesis in addition to minimizing the mean square approximation error. The task of jointly optimizing the quantizer and its subsequent linear filter was undertaken. It was shown that the quantizer that is optimized without regard to filtering nearly satisfies the necessary conditions when filtering is included. We conclude that this statically optimal design should be a good one in the majority of cases.

The optimal stochastic control of a linear system with quantized measurements subject to a quadratic cost was computed for a two-stage problem. The interpretation of the nonlinear control action leads to suggestions for suboptimal control laws. A separation theorem was derived for arbitrary nonlinear measurements under the conditions that the system is linear, the cost is quadratic, and that feedback around the nonlinear measurement device can be realized. The optimal stochastic controller is separated into two elements: a nonlinear filter for generating the conditional mean of the state and the optimal linear controller that is obtained when all uncertainties are neglected.

The truly optimal stochastic control is extremely complicated, and suboptimal algorithms must be used in most problems of any practical significance. The deficiency of a commonly used suboptimal feedback control law is alleviated by incorporating one or more future measurements into the control computations. This technique, called M-measurement-optimal feedback control, is applicable to most problems in discrete-time stochastic control.

The problem of designing linear digital compensators for closed-loop, time-invariant, linear systems when quantizers are present was discussed. A plant control loop and a communication system (quantizer with feedback) are each reduced to an equivalent Wiener filtering problem. The (approximately) optimal linear compensator must be found by an iterative solution of this equivalent problem.

7.3 Recommendations for Further Study

One of the more important subjects for study is the noisy-channel version of the topics treated in Chapters 2 and 3, such as parameter estimation, Gauss–Markov processes, and communication systems. The noiseless-channel results should be of considerable value in this endeavor.

The accuracy in the implementation of the conditional mean and conditional covariance of the state is limited only by the ingenuity and the computational capability available to the designer to find the conditional mean and covariance of the measurement vector. Any techniques that can improve the accuracy and speed of these computations will be welcome advances. Perhaps the major emphasis should be divided between relatively short vectors, such as those that might occur as a system output, and relatively large vectors (of ten to fifteen elements) that correspond to reasonable memory lengths of filters.

The analyses and simulations presented here were based on the assumption that the stochastic processes being monitored and controlled were known. It would be desirable to perform an error analysis of the Gaussian fit algorithm, for example, and determine the performance degradation that occurs when the filter is operating under the wrong assumptions. Perhaps known results from the error analyses of linear filters can be used for this task.

The quantizer with feedback is an efficient means of transmitting information, but it is likely to be inaccessible because of its remote location, and other designs should be examined in the light of the following criteria: What is the optimal fixed design for the class of spectra that are likely to be transmitted? What suboptimal computations should be performed to improve ease of mechanization and reliability? What adaptive techniques might be used to ensure nearly optimal autonomous operation? The resolution of such questions will bring the quantizer with feedback much closer to practical application.

The extension of the concept of the M-measurement-optimal feedback control to bounded control, nonquadratic criteria, and nonlinear plants is an area that may very well produce practical and efficient control strategies. There is no reason why such extensions cannot apply to the deterministic control problem as well, and numerical comparisons of these feedback control laws with the optimal control laws would be very illuminating.

These suggestions for further study are representative of the important problems in estimation and control with quantized measurements. Their solution, perhaps with some of the ideas developed here, will contribute to the understanding of systems containing other types of nonlinearities as well.

Appendix A. Power Series Approximation to the Conditional Mean and Covariance of a Quantized Gaussian Vector

The m components of a normally distributed vector are quantized and lie between the limits $\{a^i\}$ and $\{b^i\}$; that is,

$$a^i \le z^i < b^i \quad \text{or} \quad \boldsymbol{a} \le \boldsymbol{z} < \boldsymbol{b} \qquad i = 1, 2, \ldots, m. \tag{A.1}$$

The geometric center of the region A in which the vector lies is denoted by the vector

$$\gamma = \tfrac{1}{2}(\boldsymbol{b} + \boldsymbol{a}) \tag{A.2}$$

and the collection of threshold halfwidths is $\{\alpha^i\}$:

$$\boldsymbol{\alpha} = \tfrac{1}{2}(\boldsymbol{b} - \boldsymbol{a}). \tag{A.3}$$

The joint normal probability-density function for the m measurements is now expanded in a power series about the center of the region. Terms of fourth and higher order will be neglected, and Equations 2.36 and 2.37 will be used to calculate the mean and covariance of the measurement vector conditioned on $z \in A$. The power-series expansion of the normal probability-density function is

$$p(z) = p(\gamma) + \frac{\partial p}{\partial z}(\gamma)\zeta + \frac{1}{2}\zeta^T \frac{\partial^2 p}{\partial z^2}(\gamma)\zeta$$
$$+ \frac{1}{6}\sum_{i=1}^{m}\sum_{j=1}^{m}\sum_{k=1}^{m} \frac{\partial^3 p}{\partial z^i\,\partial z^j\,\partial z^k}\zeta^i\zeta^j\zeta^k + \cdots, \tag{A.4}$$

where

$$\zeta = z - \gamma, \tag{A.5}$$

$$p(z) = C \exp\left(-\tfrac{1}{2}z^T\mathbf{\Gamma}^{-1}z\right), \tag{A.6}$$

$$\frac{\partial p}{\partial z} = -p(z)z^T\mathbf{\Gamma}^{-1}, \tag{A.7}$$

$$\frac{\partial^2 p}{\partial z^2} = p(z)[\mathbf{\Gamma}^{-1}zz^T\mathbf{\Gamma}^{-1} - \mathbf{\Gamma}^{-1}], \tag{A.8}$$

and C in Equation A.6 is the normalization constant of the probability-density function.

Let the volume of the quantum region A be V:

$$V(A) = 2\alpha^1 2\alpha^2 \ldots 2\alpha^m. \tag{A.9}$$

Then the probability that z is in A is found by integrating the probability-density function over A:

$$\begin{aligned} P(z \in A) &= \int_A p(z)\,dz \\ &= \int_{-\alpha}^{\alpha} d\zeta\left(p(\gamma) + \frac{\partial p}{\partial z}\zeta + \frac{1}{2}\zeta^T\frac{\partial^2 p}{\partial z^2}\zeta + \cdots\right), \end{aligned} \tag{A.10}$$

$$P(z \in A) \approx p(\gamma)V(A)\left(1 + \frac{1}{2}\sum_{i=1}^{m} B_{ii}\frac{(\alpha^i)^2}{3}\right),$$

where the matrix $\boldsymbol{B}$ is given by

$$\boldsymbol{B} = \mathbf{\Gamma}^{-1}\gamma\gamma^T\mathbf{\Gamma}^{-1} - \mathbf{\Gamma}^{-1}. \tag{A.11}$$

The conditional mean is determined next. Consider the kth component, and integrate it over the region A:

$$\begin{aligned} \int_A z^k p(z)\,dz &= \int_A (\gamma^k + \zeta^k)p(\gamma + \zeta)\,d\zeta \\ &= \gamma^k P(z \in A) + \int_{-\alpha}^{\alpha} d\zeta\, \zeta^k p(\gamma) \\ &\quad \times \left(1 - \gamma^T\mathbf{\Gamma}^{-1}\zeta + \tfrac{1}{2}\zeta^T\boldsymbol{B}\zeta + \cdots\right). \end{aligned} \tag{A.12}$$

Only three terms are retained in the expansion of $p(z)$ since, when they are multiplied by ζ^k, their order is increased by 1. Only the second term, when multiplied by ζ^k, has an even contribution.

Thus Equation A.12 becomes

$$\begin{aligned}\int_A z^k p(z)\, dz &\approx \gamma^k P(z \in A) - \frac{V(A)}{2\alpha^k}\int_{-\alpha^k}^{\alpha^k} d\zeta^k (\gamma^T \boldsymbol{\Gamma}^{-1})^k (\zeta^k)^2 p(\gamma) \\ &= \gamma^k P(z \in A) - V(A)p(\gamma)\frac{(\alpha^k)^2}{3}(\boldsymbol{\Gamma}^{-1}\gamma)^k.\end{aligned} \tag{A.13}$$

Dividing this by $P(z \in A)$, given in Equation A.10, and retaining only third-order terms yield the approximation to the conditional mean of z:

$$E(z \mid z \in A) \approx \gamma - \boldsymbol{A}\boldsymbol{\Gamma}^{-1}\gamma, \tag{A.14}$$

where the diagonal matrix $\boldsymbol{A}$ is

$$\boldsymbol{A} = \left\{\frac{(\alpha^i)^2}{3}\delta_{ij}\right\}, \tag{A.15}$$

and δ_{ij} is the Kronecker delta.

The covariance of z conditioned on $z \in A$ is the same as ζ conditioned on $z \in A$, because the two random variables differ only by a constant. Thus,

$$\operatorname{cov}(z \mid z \in A) = \int_A \zeta\zeta^T \frac{p(\gamma + \zeta)}{P(z \in A)} d\zeta - E(\zeta \mid z \in A)E(\zeta^T \mid z \in A). \tag{A.16}$$

But the conditional mean of ζ in the rightmost term of Equation A.14 is only of second order; the rightmost term of Equation A.16 is therefore of fourth order and can be neglected.

$$\begin{aligned}\operatorname{cov}(z \mid z \in A) &\approx \int_{-\alpha}^{\alpha} \zeta\zeta^T \frac{p(\gamma)}{P(z \in A)}(1 - \gamma^T \boldsymbol{\Gamma}^{-1}\zeta + \cdots)\, d\zeta \\ &\approx \int_{-\alpha}^{\alpha} \zeta\zeta^T \frac{p(\gamma)}{V(A)p(\gamma)}\, d\zeta.\end{aligned} \tag{A.17}$$

Only the diagonal terms of the matrix in the integrand are even. Performing the integration in Equation A.17 yields

$$\operatorname{cov}(z \mid z \in A) \approx \left\{\frac{(\alpha^i)^2}{3}\delta_{ij}\right\} = \boldsymbol{A}. \tag{A.18}$$

Appendix B. Performance Estimates for the Gaussian Fit Algorithm

This appendix considers the performance of the Gaussian fit algorithm over the ensemble of time functions for which it is intended. An approximation to the ensemble mean square error is found for stationary and nonstationary data, and the analysis is applied to the three digital systems of Chapter 3.

B.1 General Form of the Performance Estimate

The Gaussian hypothesis for the Gaussian fit algorithm is assumed, so that Equations 2.56 to 2.61 describe the propagation of the first two moments. The ensemble average covariance is found by averaging the conditional covariance over all possible measurement sequences. Consider $\boldsymbol{M}_{n+1}$, the conditional covariance just prior to the measurement at t_{n+1}. From Equations 2.57 to 2.59 it is seen that Equation 2.61 can be written

$$\begin{aligned}\boldsymbol{M}_{n+1} &= \boldsymbol{\Phi}_n \boldsymbol{P}_n(\boldsymbol{M}_n)\boldsymbol{\Phi}_n^T + \boldsymbol{Q}_n \\ &\quad + \boldsymbol{\Phi}_n \boldsymbol{K}_n(\boldsymbol{M}_n) \operatorname{cov}(z_n \mid A_n, \boldsymbol{M}_n, \hat{\boldsymbol{x}}_{n|n-1}) \boldsymbol{K}_n^T(\boldsymbol{M}_n)\boldsymbol{\Phi}_n^T, \qquad \text{(B.1)}\end{aligned}$$

where the conditional mean and covariance of z_n just prior to the reading at t_n are explicitly shown in $\operatorname{cov}(z_n \mid \ldots)$. Let $\boldsymbol{M}_{n+1}^*$ be the ensemble average of $\boldsymbol{M}_{n+1}$, and formally take the ensemble expectation of both sides of Equation B.1:

$$M_{n+1}^{*} = \int dM_n p(M_n)[\Phi_n P_n(M_n)\Phi_n^T + Q_n + \Phi_n K_n(M_n)W(M_n)K_n^T(M_n)\Phi_n^T]. \tag{B.2}$$

Here $p(M_n)$ is the ensemble probability-density function for the matrix M_n, and the matrix $W(M_n)$ is

$$W(M_n) = E[\operatorname{cov}(z_n \mid A_n, \hat{x}_{n|n-1}, M_n) \mid M_n]. \tag{B.3}$$

An approximate solution to Equation B.2 can be found by expanding the right-hand side in a Taylor series about the ensemble mean M_n^* and neglecting terms of second and higher order. This will be an accurate solution when there is small probability that M_n is "far" from M_n^*, and it results in the equation

$$M_{n+1}^{*} \approx \Phi_n P_n(M_n^*)\Phi_n^T + Q_n + \Phi_n K_n(M_n^*)W(M_n^*)K_n^T(M_n^*)\Phi_n^T \tag{B.4}$$

with $M_1^* = \Phi_0 P_0 \Phi_0^T + Q_0$.

By similar reasoning the ensemble average covariance just *after* the $(n + 1)$st measurement, E_{n+1}^*, is given by

$$E_{n+1}^{*} \approx P_{n+1}(M_{n+1}^*) + K_{n+1}(M_{n+1}^*)W(M_{n+1}^*)K_{n+1}^T(M_{n+1}^*). \tag{B.5}$$

The form of $W(M_n)$ depends on the quantization scheme, and explicit forms are given below for three types of digital systems. The case of a scalar z_n is of special interest. Let $\{A^j\}$, with $j = 1, \ldots, N$, denote the N quantum intervals, and let

$$\begin{aligned} \bar{z}_n &= H_n \hat{x}_{n|n-1}, \\ \sigma_{z_n}^2 &= H_n M_n H_n^T + R_n. \end{aligned} \tag{B.6}$$

Then the expression for the scalar $W(M_n)$ becomes

$$W(M_n) = E\left[\sum_{j=1}^{N} \operatorname{cov}(z_n \mid z_n \in A^j, \bar{z}_n, \sigma_{z_n}^2) \times P(z_n \in A^j \mid \bar{z}_n, \sigma_{z_n}^2) \mid \sigma_{z_n}^2\right] \tag{B.7}$$

B.2 Pulse Code Modulation

Computation of $W(M^*)$

The estimate of the ensemble mean square error of the Gaussian fit algorithm in the pulse-code-modulation mode is determined by

Equation B.4 with $W(\boldsymbol{M})$ determined as follows. Dropping the subscripts, let the N quantum intervals A^j be defined by the boundaries $\{d^j\}$, with $j = 1, \ldots, N + 1$; that is,

$$z \in A^j \quad \text{if} \quad d^j \leq z < d^{j+1}, \qquad j = 1, 2, \ldots, N. \tag{B.8}$$

A quantizer distortion function is defined for a Gaussian variable and a particular quantizer and is given by

$$D\left(\frac{\bar{z}}{\sigma_z}, \frac{\boldsymbol{d}}{\sigma_z}\right) = \frac{1}{\sigma_z^2} \sum_{j=1}^{N} \operatorname{cov}(z \mid z \in A^j, \bar{z}, \sigma_z^2) P(z \in A^j \mid \bar{z}, \sigma_z^2), \tag{B.9}$$

where the conditional covariance and probability are evaluated under the assumption that z is normally distributed before quantization, see Equation 5.31. This distortion function is the normalized minimal mean square error in the reconstruction of the quantizer input. Substituting Equation B.9 in Equation B.7 provides

$$W(\boldsymbol{M}) = \sigma_z^2 \int D\left(\frac{\bar{z}}{\sigma_z}, \frac{\boldsymbol{d}}{\sigma_z}\right) p(\bar{z} \mid \boldsymbol{M})\, d\bar{z}. \tag{B.10}$$

An approximation to the ensemble statistics of $\bar{z}$ conditioned on $\boldsymbol{M}$ will be considered next. Let $\bar{\boldsymbol{x}}$ be the mean of the state conditioned on past data $\boldsymbol{Z}$. Then, from Equations B.6,

$$\begin{aligned} \bar{z}^2 &= \boldsymbol{H}\bar{\boldsymbol{x}}\bar{\boldsymbol{x}}^T\boldsymbol{H}^T \\ &= \boldsymbol{H}[E(\boldsymbol{x}\boldsymbol{x}^T \mid \boldsymbol{Z}) - \operatorname{cov}(\boldsymbol{x} \mid \boldsymbol{Z})]\boldsymbol{H}^T. \end{aligned} \tag{B.11}$$

The ensemble average of $\bar{z}$ is zero (or can be made zero by change of variable), so that averaging Equation B.11 over all possible measurements $\boldsymbol{Z}$ yields the variance of $\bar{z}$, or

$$\begin{aligned} \sigma_{\bar{z}}^2 &= \boldsymbol{H}(\boldsymbol{M}^a - \boldsymbol{M}^*)\boldsymbol{H}^T \\ &= (\sigma_z^a)^2 - (\sigma_z^*)^2, \end{aligned} \tag{B.12}$$

where $\boldsymbol{M}^a$ is the a priori covariance of the process, and

$$\begin{aligned} (\sigma_z^a)^2 &= \boldsymbol{H}\boldsymbol{M}^a\boldsymbol{H}^T + \boldsymbol{R}, \\ (\sigma_z^*)^2 &= \boldsymbol{H}\boldsymbol{M}^*\boldsymbol{H}^T + \boldsymbol{R}. \end{aligned} \tag{B.13}$$

Now it is assumed that the distribution of $\bar{z}$ is close to $N(0, \sigma_{\bar{z}}^2)$—which is strictly true when the measurements are linear—and the desired result is

$$W(\boldsymbol{M}_n^*) = (\sigma_{z_n}^*)^2 \int_{-\infty}^{\infty} \frac{\exp[-\frac{1}{2}(\bar{z}/\sigma_{\bar{z}_n})^2]}{(2\pi)^{1/2}\sigma_{\bar{z}_n}} D\left(\frac{\bar{z}}{\sigma_{z_n}^*}, \frac{\boldsymbol{d}}{\sigma_{z_n}^*}\right) d\bar{z}. \tag{B.14}$$

The standard deviations are found from versions of Equations B.12 or Equations B.13 with the same subscripts. Although analytical approximations can be made to Equation B.14 it can be evaluated quite easily by numerical quadrature.

The Optimal Quantizer

For three or more quantum levels the collection $\{d^i\}$ of quantizer parameters may be so chosen as to minimize the ensemble covariance $\boldsymbol{E}^*_\infty$. This was considered by the writer (1968), and it was found that for stationary processes the system mean square error is relatively insensitive to changes in $\{d^i\}$ if the quantizer is optimized for the a priori variance $(\sigma^a_z)^2$. The insensitivity is caused by two competing factors: the filtering action reduces the variance, implying that the size of the quantum intervals should be reduced but, from Equation B.12, a smaller σ^*_z means a larger variance of the conditional mean, and the quantum intervals should be increased, to insure efficient quantization when $\bar{z}$ is away from zero.

B.3 Predictive Quantization

As in the case of pulse code modulation, an estimate of the ensemble average mean square error is generated by the recursive solution of Equations B.4 and B.5 with a different form for $W(\boldsymbol{M}^*_n)$. Recall that the feedback function L_n is the mean of z_n conditioned on all past measurements. This has the effect of setting $\bar{z}$ equal to zero in the distortion function given by Equation B.9 (actually u_n is being quantized, but u_n and z_n differ only by a constant, so their covariances are the same). Consequently, $W(\boldsymbol{M}^*_n)$ for predictive quantization becomes

$$W(\boldsymbol{M}^*_n) = (\sigma^*_{z_n})^2 D\left(0, \frac{\boldsymbol{d}}{\sigma^*_{z_n}}\right). \tag{B.15}$$

The quantizer design for a stationary input may be made as follows. Temporarily assume that the quantizer is time-varying, and choose the parameters d^i at time t_n such that they are optimal for $(\sigma^*_{z_n})^2$. Now $D(0, \boldsymbol{d}/\sigma^*_{z_n})$ becomes $D_g(N)$, the minimal distortion for a unit-variance, zero-mean, Gaussian variable, and it is a function only of N, the number of quantum levels; see Max, 1960, for tabulations of $\{d^i\}$ and $D_g(N)$. With this time-varying quantizer $W(\boldsymbol{M}^*_n)$ becomes

$$W(\boldsymbol{M}^*_n) = (\sigma^*_{z_n})^2 D_g(N). \tag{B.16}$$

As n approaches infinity, the ensemble covariance and thus the quantizer parameters approach a constant. This final quantizer minimizes the ensemble covariance, because using any distortion other than $D_g(N)$ yields a larger solution to the Riccati equation, Equation B.4.

B.4 Data-Compression

The $W(\boldsymbol{M})$ function for the data-compression system is identical with that in predictive quantization, except that there is only one quantum interval (of halfwidth α) in the distortion function. Equation B.15 becomes

$$W(\boldsymbol{M}_n^*) = (\sigma_{z_n}^*)^2 \left[\int_{-\alpha/\sigma_{z_n}^*}^{\alpha/\sigma_{z_n}^*} \frac{\exp(-\frac{1}{2}u^2)\,du}{(2\pi)^{1/2}} - 2\left(\frac{\alpha}{\sigma_{z_n}^*}\right) \frac{\exp[-\frac{1}{2}(\alpha/\sigma_{z_n}^*)^2]}{(2\pi)^{1/2}} \right]. \tag{B.17}$$

Equations B.4, B.5, and B.17 describe an approximation to the ensemble covariance.

The sample-compression ratio, CR, is a common figure of merit for data-compression systems; it is the ratio of the number of input samples to the average number of transmitted samples. For the system considered here it is

$$\mathrm{CR}(\boldsymbol{M}_\infty^*) = \left(1 - \int_{-\alpha/\sigma_{z_\infty}^*}^{\alpha/\sigma_{z_\infty}^*} \frac{\exp(-\frac{1}{2}u^2)\,du}{(2\pi)^{1/2}}\right)^{-1}. \tag{B.18}$$

Appendix C. Sensitivities of Mean Square Estimation Error with Respect to Quantizer Parameters

This appendix contains the derivation of the partial derivatives of the mean square estimation error (MSEE) with respect to the quantizer parameters.

C.1 Change in MSEE due to Changes in Output Autocorrelation

Equation 4.45 is repeated here:

$$\delta\overline{e^2} = \boldsymbol{K}\mathsf{k}^2\delta\mathsf{Y}\boldsymbol{K}^T. \tag{C.1}$$

The $M \times M$ matrix $\delta\mathsf{Y}$ is composed of the elements

$$\delta\mathsf{Y}_{ij} = \delta\left(\frac{\phi_{yy}(i-j)}{\mathsf{k}^2}\right). \tag{C.2}$$

This matrix has only M independent elements. As in Equation 4.46, let

$$b(m) = \frac{\phi_{yy}(m)}{\mathsf{k}^2}. \tag{C.3}$$

Now write $\delta\mathsf{Y}$ as the sum of three matrices:

$$\delta\mathsf{Y} = \boldsymbol{D} + \boldsymbol{C} + \boldsymbol{C}^T, \tag{C.4}$$

where

$$\boldsymbol{D} = \begin{pmatrix} \delta b(0) & & \mathbf{0} \\ & \delta b(0) & \\ & & \ddots \\ \mathbf{0} & & \delta b(0) \end{pmatrix},$$

$$\boldsymbol{C} = \begin{pmatrix} 0 & & & & \\ \delta b(1) & 0 & & \mathbf{0} & \\ \delta b(2) & \delta b(1) & 0 & & \\ & & & \ddots & 0 \\ \delta b(M-1) & \cdots\cdots & & \delta b(1) & 0 \end{pmatrix}. \tag{C.5}$$

Substituting Equations C.4 and C.5 in Equation C.1 yields

$$\begin{aligned} \delta\overline{e^2} &= \mathsf{k}^2 \boldsymbol{KDK}^T + 2\mathsf{k}^2 \boldsymbol{KCK}^T \\ &= \mathsf{k}^2 \sum_{i=1}^{M} K_i^2 \delta b(0) + 2\mathsf{k}^2 \sum_{i=2}^{M} \sum_{j=1}^{i-1} K_i K_j \delta b(i-j). \end{aligned} \tag{C.6}$$

By making a change of subscripts in the second term this term may be rewritten

$$\begin{aligned} 2\mathsf{k}^2 \sum_{i=2}^{M} \sum_{j=1}^{i-1} K_i K_j \delta b(i-j) &= 2\mathsf{k}^2 \sum_{i=2}^{M} \sum_{m=1}^{i-1} K_i K_{i-m} \delta b(m) \\ &= 2\mathsf{k}^2 \sum_{m=1}^{M-1} \delta b(m) \sum_{j=1}^{M-m} K_j K_{j+m}. \end{aligned} \tag{C.7}$$

The change in mean square estimation error due to changes in the b variables, Equation C.6, now becomes

$$\delta\overline{e^2} = \left(\mathsf{k}^2 \sum_{i=1}^{M} K_i^2\right) \delta b(0) + \sum_{m=1}^{M-1} \left(2\mathsf{k}^2 \sum_{j=1}^{M-m} K_j K_{j+m}\right) \delta b(m). \tag{C.8}$$

C.2 Partial Derivatives of $b(m)$ with Respect to $\{d^n\}$

The partial derivatives of the $b(m)$ variables with respect to the quantizer input parameters d^n will be treated first. From Equation C.3 we have

$$\frac{\partial b(m)}{\partial d^n} = \frac{1}{\mathsf{k}^2}\left(\frac{\partial \phi_{yy}(m)}{\partial d^n} - \frac{2\phi_{yy}(m)}{\mathsf{k}} \frac{\partial \mathsf{k}}{\partial d^n}\right). \tag{C.9}$$

The first term here is found by using the expressions from Equations 4.1 and 4.2, when m is nonzero:

$$\frac{\partial \phi_{yy}(m)}{\partial d^n} = \frac{\partial}{\partial d^n} \sum_{r=1}^{N} \sum_{s=1}^{N} y^r y^s \int_{d^r}^{d^{r+1}} d\zeta_1 \int_{d^s}^{d^{s+1}} d\zeta_2 p_{z_{j+m}, z_j}(\zeta_1, \zeta_2), \tag{C.10}$$

where $p_{z_{j+m}, z_j}(\zeta_1, \zeta_2)$ is the joint probability-density function for the quantizer input samples at times t_{j+m} and t_j.

Continuing with Equation C.10 gives

$$\begin{aligned} \frac{\partial \phi_{yy}(m)}{\partial d^n} &= 2 \sum_{s=1}^{N} y^{n-1} y^s \int_{d^s}^{d^{s+1}} p_{z_{j+m}|z_j}(\zeta_2, d^n)\, d\zeta_2 p_{z_j}(d^n) \\ &\quad - 2 \sum_{s=1}^{N} y^n y^s \int_{d^s}^{d^{s+1}} p_{z_{j+m}|z_j}(\zeta_2, d^n)\, d\zeta_2 p_{z_j}(d^n) \\ &= 2(y^{n-1} - y^n) p_{z_j}(d^n) \sum_{s=1}^{N} y^s P(z_{j+m} \in A^s \mid z_j = d^n) \\ &= 2(y^{n-1} - y^n) p_{z_j}(d^n) E(y_{j+m} \mid z_j = d^n). \end{aligned} \tag{C.11}$$

When m is zero, the derivative is given by

$$\begin{aligned} \frac{\partial \phi_{yy}(0)}{\partial d^n} &= \frac{\partial}{\partial d^n} \sum_{s=1}^{N} (y^s)^2 \int_{d^s}^{d^{s+1}} p_{z_j}(\zeta)\, d\zeta \\ &= (y^{n-1})^2 p_{z_j}(d^n) - (y^n)^2 p_{z_j}(d^n) \\ &= [(y^{n-1})^2 - (y^n)^2] p_{z_j}(d^n). \end{aligned} \tag{C.12}$$

The partial derivatives of k with respect to d^n are determined next. Equation 4.17 gives

$$\begin{aligned} \frac{\partial \mathrm{k}}{\partial d^n} &= \frac{\partial}{\partial d^n} \frac{1}{\sigma_z^2} \sum_{s=1}^{N} y^s \int_{d^s}^{d^{s+1}} \zeta p_z(\zeta)\, d\zeta \\ &= \frac{1}{\sigma_z^2} [y^{n-1} d^n p_z(d^n) - y^n d^n p_z(d^n)] \\ &= \frac{1}{\sigma_z^2} (y^{n-1} - y^n) d^n p_z(d^n). \end{aligned} \tag{C.13}$$

Substituting Equations C.11, C.12, and C.13 in Equation C.9 yields the desired form of the partial derivative of $b(m)$ with respect to d^n:

$$\frac{\partial b(m)}{\partial d^n} = \begin{cases} \dfrac{2}{\mathsf{k}^2}(y^{n-1} - y^n)p_z(d^n) \\ \qquad \times \left(\dfrac{y^{n-1} + y^n}{2} - \dfrac{\phi_{yy}(0)}{\mathsf{k}\sigma_z^2} d^n\right), & m = 0, \\ \dfrac{2}{\mathsf{k}^2}(y^{n-1} - y^n)p_z(d^n) \\ \qquad \times \left(E(y_{j+m} \mid z_j = d^n) - \dfrac{\phi_{yy}(m)}{\mathsf{k}\sigma_z^2} d^n\right), & m \neq 0. \end{cases} \quad \text{(C.14)}$$

C.3 Partial Derivatives of *b(m)* with respect to $\{y^n\}$

The partial derivatives of the $b(m)$ variables with respect to the quantizer output parameters y^n are found from Equation C.3:

$$\frac{\partial b(m)}{\partial y^n} = \frac{1}{\mathsf{k}^2}\left(\frac{\partial \phi_{yy}(m)}{\partial y^n} - \frac{2\phi_{yy}(m)}{\mathsf{k}} \frac{\partial \mathsf{k}}{\partial y^n}\right). \quad \text{(C.15)}$$

The first term for nonzero m is found by using the expressions from Equation 4.1:

$$\begin{aligned} \frac{\partial \phi_{yy}(m)}{\partial y^n} &= \frac{\partial}{\partial y^n} \sum_{r=1}^{N} \sum_{s=1}^{N} y^r y^s P(z_{j+m} \in A^r, z_j \in A^s) \\ &= \sum_{s=1}^{N} y^s P(z_{j+m} \in A^n, z_j \in A^s) \\ &\quad + \sum_{r=1}^{N} y^r P(z_{j+m} \in A^r, z_j \in A^n) \\ &= 2\left(\sum_{r=1}^{N} y^r P(z_{j+m} \in A^r \mid z_j \in A^n)\right) P(z_j \in A^n) \\ &= 2P(z_j \in A^n) E(y_{j+m} \mid y_j = y^n). \end{aligned} \quad \text{(C.16)}$$

When m is zero, the expression equivalent to Equation C.16 is

$$\frac{\partial \phi_{yy}(0)}{\partial y^n} = \frac{\partial}{\partial y^n} \sum_{s=1}^{N} (y^s)^2 P(z \in A^s) = 2y^n P(z \in A^s) \quad \text{(C.17)}$$

and it is seen that Equation C.16 can also treat the case of $m = 0$.

The partial derivative of k with respect to y^n is calculated by using Equation 4.17:

$$\frac{\partial \mathsf{k}}{\partial y^n} = \frac{\partial}{\partial y^n} \frac{1}{\sigma_z^2} \sum_{r=1}^{N} y^r E(z \mid z \in A^r) P(z \in A^r)$$

$$= \frac{1}{\sigma_z^2} E(z \mid z \in A^n) P(z \in A^n). \tag{C.18}$$

Equations C.16 and C.18 are substituted in Equation C.15 to provide the partial derivative of $b(m)$ with respect to y^n:

$$\frac{\partial b(m)}{\partial y^n} - \frac{2}{\mathsf{k}^2} P(z \in A^n) \left(E(y_{j+m} \mid y_j = y^n) - \frac{\phi_{yy}(m)}{\mathsf{k}\sigma_z^2} E(z \mid z \in A^n) \right). \tag{C.19}$$

Appendix D. A Proof of a Separation Theorem for Nonlinear Measurements

D.1 The Distribution of the Error in the Estimate

This, the first part of the proof, shows that the conditional probability-density function of the error in the estimate is not a function of past control actions. Equation 5.39 for the residual, Equation 5.40 for the prediction error, and Equation 5.41 for the measurement equation with feedback are repeated here for convenience:

$$\boldsymbol{r}_{k|k-1} = \boldsymbol{H}_k \boldsymbol{e}_{k|k-1}, \tag{D.1}$$

$$\boldsymbol{e}_{k|k-1} = \boldsymbol{\Phi}_{k-1} \boldsymbol{e}_{k-1|k-1} + \boldsymbol{w}_{k-1}, \tag{D.2}$$

$$\boldsymbol{z}_k = \boldsymbol{h}_k(\boldsymbol{r}_{k|k-1}, \boldsymbol{v}_k). \tag{D.3}$$

Any function that does not include the control variables (but it may include past measurements) can be added to the residual or be multiplied by the residual, and the following arguments are still valid.

At time t_1 *prior* to the processing of the measurement z_1 the distributions of the error in the estimate, $\boldsymbol{e}_{1|0}$, the residual $\boldsymbol{r}_{1|0}$, and the measurement z_1 are all determined solely by the distributions of $\boldsymbol{e}_{0|0}$, $\boldsymbol{w}_0$, and Equations D.1, D.2, and D.3, and these distributions are therefore not influenced by $\boldsymbol{u}_0$. Now we shall show that the distribution of the error in the estimate just *after* the first measurement is not influenced by $\boldsymbol{u}_0$, either. Instead of considering the estimation error, though, it is more convenient to consider the prediction error $\boldsymbol{e}_{1|0}$ conditioned on the measurement z_1 (the conditional distribution is the same as that for the estimation error except for a bias). This conditional density function in its most general form is $p(\boldsymbol{e}_{1|0} \mid z_1, \boldsymbol{Z}_0, \boldsymbol{u}_0)$. Applying Bayes' rule, we have

$$p(\boldsymbol{e}_{1|0} \mid z_1, \boldsymbol{Z}_0, \boldsymbol{u}_0) = \frac{p(z_1 \mid \boldsymbol{e}_{1|0}, \boldsymbol{Z}_0, \boldsymbol{u}_0)\, p(\boldsymbol{e}_{1|0} \mid \boldsymbol{Z}_0, \boldsymbol{u}_0)}{p(z_1 \mid \boldsymbol{Z}_0, \boldsymbol{u}_0)}. \tag{D.4}$$

Each of the three conditional density functions on the right-hand side of this equation is independent of $\boldsymbol{u}_0$: the function $p(z_1 \mid \boldsymbol{e}_{1|0}, \boldsymbol{Z}_0, \boldsymbol{u}_0)$, because of Equations D.1 and D.3; the function $p(\boldsymbol{e}_{1|0} \mid \boldsymbol{Z}_0, \boldsymbol{u}_0)$, because of Equation D.2; and the function $p(z_1 \mid \boldsymbol{Z}_0, \boldsymbol{u}_0)$, because of Equations D.1, D.2, and D.3. Thus the distribution of the error in the estimate *after* the first measurement is independent of $\boldsymbol{u}_0$, and so are all subsequent error distributions. The same is true of all later control actions, so that the distribution of the error in the estimate at each time never depends on the control.

D.2 The Optimal Control Sequence

Dynamic programming is used to find the optimal control sequence. Proceeding to the last stage, the optimal choice of $\boldsymbol{u}_N$ is that which minimizes the performance criterion,

$$J_N = E(\boldsymbol{x}_N^T \boldsymbol{A}_N \boldsymbol{x}_N + \boldsymbol{u}_N^T \boldsymbol{B}_N \boldsymbol{u}_N + \boldsymbol{x}_{N+1}^T \boldsymbol{A}_{N+1} \boldsymbol{x}_{N+1} \mid \boldsymbol{Z}_N, \boldsymbol{U}_N) \tag{D.5}$$

subject to the state-equation constraint, Equation 5.36. The best choice of $\boldsymbol{u}_N$ by direct minimization is

$$\boldsymbol{u}_N^o = -(\boldsymbol{B}_N + \boldsymbol{G}_N^T \boldsymbol{S}_{N+1} \boldsymbol{G}_N)^{-1} \boldsymbol{G}_N^T \boldsymbol{S}_{N+1} \boldsymbol{\Phi}_N E(\boldsymbol{x}_N \mid \boldsymbol{Z}_N, \boldsymbol{U}_{N-1}), \tag{D.6}$$

where we have made the substitution

$$\boldsymbol{S}_{N+1} = \boldsymbol{A}_{N+1}.$$

From Equation 5.20 the cost of completing the process from time t_{N-1} with the use of *any* $\boldsymbol{u}_{N-1}$ but the *optimal* $\boldsymbol{u}_N^o$, can be written

$$\begin{aligned} J_{N-1} = {} & E(\boldsymbol{x}_{N-1}^T \boldsymbol{A}_{N-1} \boldsymbol{x}_{N-1} + \boldsymbol{u}_{N-1}^T \boldsymbol{B}_{N-1} \boldsymbol{u}_{N-1} \\ & + \boldsymbol{x}_N^T \boldsymbol{S}_N \boldsymbol{x}_N \mid \boldsymbol{Z}_{N-1}, \boldsymbol{U}_{N-1}) \\ & + \operatorname{tr}[E(\boldsymbol{C}_N \boldsymbol{E}_{N|N} \mid \boldsymbol{Z}_{N-1}, \boldsymbol{U}_{N-1})] + \text{const}, \end{aligned} \tag{D.7}$$

where

$$\boldsymbol{S}_N = \boldsymbol{A}_N + \boldsymbol{\Phi}_N^T \boldsymbol{S}_{N+1} \boldsymbol{\Phi}_N - \boldsymbol{C}_N, \tag{D.8}$$

$$\boldsymbol{C}_N = \boldsymbol{\Phi}_N^T \boldsymbol{S}_{N+1} \boldsymbol{G}_N (\boldsymbol{B}_N + \boldsymbol{G}_N^T \boldsymbol{S}_{N+1} \boldsymbol{G}_N)^{-1} \boldsymbol{G}_N^T \boldsymbol{S}_{N+1} \boldsymbol{\Phi}_N, \tag{D.9}$$

$$\boldsymbol{E}_{N|N} = E(\boldsymbol{e}_{N|N} \boldsymbol{e}_{N|N}^T \mid \boldsymbol{Z}_N, \boldsymbol{U}_{N-1}). \tag{D.10}$$

The optimal choice of $\boldsymbol{u}_{N-1}$ minimizes Equation D.7. But the first part of this proof showed that the distributions of both the estimation error and the measurements are independent of control action. The second term in Equation D.7 may thus be ignored, and an equivalent performance criterion is given by

$$J'_{N-1} = E(\boldsymbol{x}_{N-1}^T \boldsymbol{A}_{N-1} \boldsymbol{x}_{N-1} + \boldsymbol{u}_{N-1}^T \boldsymbol{B}_{N-1} \boldsymbol{u}_{N-1} + \boldsymbol{x}_N^T \boldsymbol{S}_N \boldsymbol{x}_N \mid \boldsymbol{Z}_{N-1}, \boldsymbol{U}_{N-1}). \tag{D.11}$$

Since this is precisely the form of J_N in Equation D.5, with all subscripts decreased by 1, it follows that $\boldsymbol{u}_{N-1}^o$ is determined by Equation D.6 with all subscripts decreased by 1. The remaining controls are found by induction, with the result that

$$\boldsymbol{u}_i^o = -(\boldsymbol{B}_i + \boldsymbol{G}_i^T \boldsymbol{S}_{i+1} \boldsymbol{G}_i)^{-1} \boldsymbol{G}_i^T \boldsymbol{S}_{i+1} \boldsymbol{\Phi}_i E(\boldsymbol{x}_i \mid \boldsymbol{Z}_i, \boldsymbol{U}_{i-1}), \tag{D.12}$$

$$\boldsymbol{S}_i = \boldsymbol{A}_i + \boldsymbol{\Phi}_i^T [\boldsymbol{S}_{i+1} - \boldsymbol{S}_{i+1} \boldsymbol{G}_i (\boldsymbol{B}_i + \boldsymbol{G}_i^T \boldsymbol{S}_{i+1} \boldsymbol{G}_i)^{-1} \boldsymbol{G}_i^T \boldsymbol{S}_{i+1}] \boldsymbol{\Phi}_i,$$

$$\boldsymbol{S}_{N+1} = \boldsymbol{A}_{N+1}. \tag{D.13}$$

This control law is identical with Equation 5.22 for the linear-measurement case in which the optimal stochastic controller is separated into two elements, a filter for generating the conditional mean of the state vector and a linear control law that is optimal when all uncertainties are neglected.

Appendix E. Derivation of the One-Measurement Cost Function

Assume that measurements and control actions have occurred up to time t_k. The control $\boldsymbol{u}_k$ is determined by allowing for the measurement at time t_{k+1} and assuming an open-loop program from t_{k+1} on. The cost of completing this process for a linear system and a quadratic cost function is, from Equation 6.19,

$$\begin{aligned} J_{OM,k} = E[&\boldsymbol{x}_k^T \boldsymbol{A}_k \boldsymbol{x}_k + \boldsymbol{u}_k^T \boldsymbol{B}_k \boldsymbol{u}_k \\ &+ E(J_{OL,k+1}^o \mid \boldsymbol{Z}_k, \boldsymbol{U}_k) \mid \boldsymbol{Z}_k, \boldsymbol{U}_{k-1}], \end{aligned} \tag{E.1}$$

where $J_{OL,k+1}^o$ is the minimal cost of completing the open-loop portion of the program; it is found from Equation 6.12 by replacing k with $k+1$:

$$J_{OL,k+1}^o = \hat{\boldsymbol{x}}_{k+1|k+1}^T \boldsymbol{S}_{k+1} \hat{\boldsymbol{x}}_{k+1|k+1} + \sum_{i=k+1}^{N+1} \operatorname{tr}(\boldsymbol{A}_i \boldsymbol{E}_{i|k+1}), \tag{E.2}$$

where

$$\begin{aligned} \hat{\boldsymbol{x}}_{k+1|k+1} &= E(\boldsymbol{x}_{k+1} \mid \boldsymbol{Z}_{k+1}, \boldsymbol{U}_k), \\ \boldsymbol{E}_{i|k+1} &= E[(\boldsymbol{x}_i - \hat{\boldsymbol{x}}_{i|k+1})(\boldsymbol{x}_i - \hat{\boldsymbol{x}}_{i|k+1})^T \mid \boldsymbol{Z}_{k+1}, \boldsymbol{U}_{i-1}], \\ \boldsymbol{S}_i &= \boldsymbol{A}_i + \boldsymbol{\Phi}_i^T[\boldsymbol{S}_{i+1} - \boldsymbol{S}_{i+1}\boldsymbol{G}_i(\boldsymbol{B}_i + \boldsymbol{G}_i^T \boldsymbol{S}_{i+1} \boldsymbol{G}_i)^{-1} \boldsymbol{G}_i^T \boldsymbol{S}_{i+1}]\boldsymbol{\Phi}_i, \\ \boldsymbol{S}_{N+1} &= \boldsymbol{A}_{N+1}. \end{aligned} \tag{E.3}$$

The last term in Equation E.2 may also be written

$$\sum_{i=k+1}^{N+1} \operatorname{tr}(A_i E_{i|k+1}) = \sum_{i=k+1}^{N+1} E(e_{i|k+1}^T A_i e_{i|k+1} \mid Z_{k+1}, U_{i-1}). \tag{E.4}$$

The error propagates through the system during this open-loop control phase as indicated in Equation 6.3:

$$e_{i+1|k+1} = \Phi_i e_{i|k+1} + w_i \tag{E.5}$$

$$= \Phi_{i+1,k+1} e_{k+1|k+1} + \sum_{j=k+1}^{i} \Phi_{i+1,j+1} w_j, \tag{E.6}$$

where $\Phi_{i+1,j+1} = \Phi_i \Phi_{i-1} \ldots \Phi_{j+1}$.

Equation E.6 may be substituted in Equation E.4. All cross-product terms drop out, because the noise vectors are independent of the estimation error $e_{k+1|k+1}$ and of each other. The terms involving the noise covariances are additive constants. Thus Equation E.4 becomes

$$\begin{aligned}
&\sum_{i=k+1}^{N+1} \operatorname{tr}(A_i E_{i|k+1}) \\
&\quad = \sum_{i=k+1}^{N+1} E(e_{k+1|k+1}^T \Phi_{i,k+1}^T A_i \Phi_{i,k+1} e_{k+1|k+1} \mid Z_{k+1}, U_{i-1}) \\
&\qquad + \text{const} \\
&\quad = \operatorname{tr}\left(E_{k+1|k+1} \sum_{i=k+1}^{N+1} \Phi_{i,k+1}^T A_i \Phi_{i,k+1} \right) + \text{const.}
\end{aligned} \tag{E.7}$$

The summation of matrices in Equation E.7 depends only on the system and the cost function. Define this matrix as F_{k+1}:

$$F_{k+1} = \sum_{i=k+1}^{N+1} \Phi_{i,k+1}^T A_i \Phi_{i,k+1}$$

$$= \Phi_{k+1}^T F_{k+2} \Phi_{k+1} + A_{k+1}, \tag{E.8}$$

$$F_{N+1} = A_{N+1}. \tag{E.9}$$

An equivalent expression for the optimal open-loop cost $J_{OL,k+1}^o$ may be found by substituting Equation E.8 in Equation E.7 and, in turn, substituting the result in Equation E.2:

$$
\begin{aligned}
J^o_{OL,k+1} &= \hat{\boldsymbol{x}}^T_{k+1|k+1}\boldsymbol{S}_{k+1}\hat{\boldsymbol{x}}_{k+1|k+1} \\
&\quad + \mathrm{tr}\,(\boldsymbol{E}_{k+1|k+1}\boldsymbol{F}_{k+1}) + \text{const} \\
&= E(\boldsymbol{x}^T_{k+1}\boldsymbol{S}_{k+1}\boldsymbol{x}_{k+1} \mid \boldsymbol{Z}_{k+1}, \boldsymbol{U}_k) \\
&\quad + \mathrm{tr}\,[\boldsymbol{E}_{k+1|k+1}(\boldsymbol{F}_{k+1} - \boldsymbol{S}_{k+1})] + \text{const.}
\end{aligned} \tag{E.10}
$$

The second of these two expressions for the cost is averaged over all measurements at t_{k+1} and substituted in Equation E.1, to give $J_{OM,k}$, the one-measurement cost function:

$$
\begin{aligned}
J_{OM,k} &= E(\boldsymbol{x}^T_k\boldsymbol{A}_k\boldsymbol{x}_k + \boldsymbol{u}^T_k\boldsymbol{B}_k\boldsymbol{u}_k + \boldsymbol{x}^T_{k+1}\boldsymbol{S}_{k+1}\boldsymbol{x}_{k+1} \mid \boldsymbol{Z}_k, \boldsymbol{U}_k) \\
&\quad + \mathrm{tr}\,\{E[\boldsymbol{E}_{k+1|k+1}(\boldsymbol{F}_{k+1} - \boldsymbol{S}_{k+1}) \mid \boldsymbol{Z}_k, \boldsymbol{U}_k]\} + \text{const.}
\end{aligned} \tag{E.11}
$$

As a final point it will be shown that the weighting matrix $\boldsymbol{F}_{k+1} - \boldsymbol{S}_{k+1}$ for $\boldsymbol{E}_{k+1|k+1}$ in Equation E.11 is at least positive semidefinite (if this matrix were indefinite or negative definite, the control would minimize the cost by degrading, not improving, the measurements). To show this we shall draw upon the duality between optimal control of deterministic linear systems subject to quadratic costs and state-variable estimation (Kalman, 1960). Let us make the following interpretation of the matrices in Equation E.3:

$\boldsymbol{A}_{N+1} = \boldsymbol{S}_{N+1}$ = covariance of state vector at time t_N prior to processing of the measurement at t_N,
$\boldsymbol{A}_i$ = covariance matrix of (white) process noise at t_i,
$\boldsymbol{G}^T_i$ = measurement matrix at t_i,
$\boldsymbol{B}_i$ = covariance of (white) observation noise at t_i,
$\boldsymbol{S}_{i+1}$ = covariance of prediction error at time t_i,
$\boldsymbol{\Phi}^T_i$ = transition matrix of a system from time t_i to t_{i-1}.

Then Equation (E.3) describes the evolution of the prediction-error covariance matrix as measurements are processed; the time index is decreasing as more measurements are processed. By Equation E.9 the matrix $\boldsymbol{F}_{i+1}$ is equivalent to the error covariance at time t_i *without* the benefit of observations. The matrix $\boldsymbol{F}_{i+1} - \boldsymbol{S}_{i+1}$ represents the improvement in covariance due to measurements and is always at least positive semidefinite.

References

Aoki, M., *Optimization of Stochastic Systems*, Academic Press, New York, 1967.

Balakrishnan, A., "An Adaptive Nonlinear Data Predictor," *Proc. Natl. Telemetry Conf.* (1962).

Bass, R., and L. Schwartz, "Extensions to Multichannel Nonlinear Filtering," *Hughes Report SSD 60220R*, February 1966.

Bellman, R., *Adaptive Control Processes*, Princeton Univ. Press, Princeton, New Jersey.

Bello, P., R. Lincoln, and H. Gish, "Statistical Delta Modulation," *Proc. IEEE*, **55**, 308–319 (March 1967).

Bennett, W., "Spectra of quantized signals," *Bell System Tech. J.*, **27**, 446–472 (July 1948).

Bertram, J. E., "The Effect of Quantization in Sampled-Feedback Systems," *Trans. Am. Inst. Elec. Engrs.*, Pt. II (Appl. Ind.), **77**, 177–182 (1958).

Bryson, A. E., and Y. C. Ho, *Applied Optimal Control*, Blaisdell Publishing Company, Waltham, Massachusetts, 1969.

Bucy, R., "Nonlinear Filtering Theory," *IEEE Trans. Automatic Control*, **AC-10**, 198 (1965).

Curry, R. E., "Estimation and Control with Quantized Measurements," Ph.D. Thesis, Department of Aeronautics and Astronautics, Massachusetts Institute of Technology, Cambridge, Massachusetts, May 1968.

Davenport, W. B., and W. L. Root, *An Introduction to the Theory of Random Signals and Noise*, McGraw-Hill Book Company, Inc., New York, 1958.

Davisson, L., "Theory of Adaptive Data Compression," in *Advances in Communication Systems*, A. Balakrishnan (Ed.), Academic Press, New York, 1966.

Davisson, L., "An Approximate Theory of Prediction for Data Compression," *IEEE Trans. Inform. Theory*, **IT-13**, 274–278 (April 1967).

Davisson, L., "The Theoretical Analysis of Data Compression Techniques," *Proc. IEEE*, **56**, 176–186 (February 1968).

Deyst, J., "Optimal Control in the Presence of Measurement Uncertainties," Sc.D. Thesis, Department of Aeronautics and Astronautics, Massachusetts Institute of Technology, Cambridge, Massachusetts, January 1967.

Doob, J., *Stochastic Processes*, John Wiley & Sons, Inc., New York, 1953.

Dreyfus, S., "Some Types of Optimal Control of Stochastic Systems," *J. Soc. Ind. Appl. Math.*, Ser. A, **2,** 120–134 (1964).

Dreyfus, S., *Dynamic Programming and the Calculus of Variations*, Academic Press, New York, 1965.

Ehrman, L., "Analysis of Some Redundancy Removal Bandwidth Compression Techniques," *Proc. IEEE*, **55,** 278–287 (March 1967).

Fel'dbaum, A., "Dual Control Theory, I-IV," *Automation and Remote Control*: **21,** 874, 1033 (1960); **22,** 1, 109 (1961).

Fine, T., "Properties of an Optimum Digital System and Applications," *IEEE Trans. Inform. Theory*, **IT-10,** 287–296 (1964).

Fisher, J., "Conditional Probability Density Functions and Optimal Nonlinear Estimation," Ph.D. Dissertation, Univ. Calif., Los Angeles, California, 1966.

Fraser, D., "A New Technique for the Optimal Smoothing of Data," Sc.D. Thesis, Department of Aeronautics and Astronautics, Massachusetts Institute of Technology, Cambridge, Massachusetts, January 1967.

Gelb, A., and W. Vander Velde, *Multiple Input Describing Functions and Nonlinear System Design*, McGraw-Hill Book Company, Inc., New York, 1968.

Gish, H., "Optimum Quantization of Random Sequences," *Rept. 529*, Division of Engineering and Applied Physics, Harvard Univ., May 1967.

Graham, D. and D. McRuer, *Analysis of Nonlinear Control Systems*, John Wiley & Sons, Inc., New York, 1961.

Grenander, U., and G. Szego, *Toeplitz Forms and Their Applications*, Univ. Calif. Press, Berkeley, California, 1958.

Gunckel, T., and G. Franklin, "A General Solution for Linear, Sampled Data Control," *Trans. ASME, Ser. D, J. Basic Eng.*, **85,** 197–203 (1963).

Gupta, S. S., "Bibliography on the Multivariate Normal Integrals and Related Topics," *Ann. Math. Statist.*, **34,** 829–838 (1963).

Ho, Y. C., and R. C. K. Lee, "A Bayesian Approach to Problems in Stochastic Estimation and Control," *Proc. Joint Automatic Control Conference*, 382–387 (1964).

Huang, J., and P. Schultheiss, "Block Quantization of Correlated Gaussian Random Variables," *IEEE Trans. Commun. Systems*, Sept., 289–296 (1962).

IEEE (Inst. Elec. Electron. Engrs.) *Proc. IEEE*, Special Issue on Redundancy Reduction, **55** (March 1967).

Irwin, J., and J. O'Neal, "The Design of Optimum DPCM (Differential Pulse Code Modulation) Encoding Systems Via the Kalman Predictor," *JACC Preprints*, 130–136, June 1968.

Jazwinski, A., "Filtering for Nonlinear Dynamical Systems," *IEEE Trans. Automatic Control*, **AC-11**, 765 (1966).

Johnson, G. W., "Upper Bound on Dynamic Quantization Error in Digital Control Systems Via the Direct Method of Liapunov," *IEEE Trans. Automatic Control*, **AC-10**, 439–448 (1965).

Joseph, P., and J. Tou, "On Linear Control Theory," *Trans. Am. Inst. Elec. Engrs.*, Pt. II (Appl. Ind.), **80,** 18 (1961).

Kalman, R., "A New Approach to Linear Filtering and Prediction Problems," *Trans. ASME, Ser. D, J. Basic Eng.*, **27,** 35–45 (March 1960).

Kalman, R., and R. Bucy, "New Results in Linear Filtering and Prediction Theory," *Trans. ASME, Ser. D, J. Basic Eng.*, **28,** 95–108 (March 1961).

Kellog, W., "Information Rates in Sampling and Quantization," *IEEE Trans. Inform. Theory*, **IT-13,** 506–511 (1967).

Klerer, M., and G. Korn, *Digital Computer User's Handbook*, Chap. 2.5, McGraw-Hill Book Company, Inc., New York, 1967.

Korsak, A., "Linear Filtering and Prediction with Quantized Measurements," *Mem. 10, Project ESU 5830*, Standford Research Institute, Menlo Park, California, February 1967.

Kosyakin, A. A., "Taking the Effect of Quantization by Level into Account in the Statistical Analysis of Closed-Loop Digital Automatic Systems," *Automation and Remote Control*, **27**, 799–805 (1966).

Kushner, H., "On the Differential Equations Satisfied by Conditional Probability Densities of Markov Processes, with Applications," *J. Soc. Ind. Appl. Math., Ser. A*, **2**, 106–119 (1964).

Lanning, J. H., and R. H. Battin, *Random Processes in Automatic Control*, McGraw-Hill Book Company, Inc., New York, 1956.

Lee, R. C. K., *Optimal Estimation, Identification, and Control*, MIT Press, Cambridge, Massachusetts, 1964.

Max, J., "Quantizing for Minimum Distortion," *IRE Trans. Inform. Theory*, **IT-6**, 7–12 (March 1960).

(a) Meier, L., A. Korsak, and R. Larson, "Effect of Data Quantization on Tracker Performance," *Tech. Mem. 3, Project 6642*, Stanford Research Institute, Menlo Park, California, October 1967.

(b) Meier, L., J. Peschon, and R. Dressler, "Optimal Control of Measurement Subsystems," *IEEE Trans. Automatic Control*, **AC-12**, 528–536 (1967).

Mowery, V., "Least Squares Differential-Correction Estimation in Nonlinear Problems," *IEEE Trans. Automatic Control*, **AC-10**, 399–407 (1965).

O'Neal, J., "Predictive Quantizing Systems (Differential Pulse Code Modulation) for the Transmission of Television Signals," *Bell System Tech. J.*, **45**, 689–720 (1966).

Papoulis, A., *Probability, Random Variables, and Stochastic Processes*, McGraw-Hill Book Company, Inc., New York, 1965.

Quagliata, L. J., private communication, March 1968.

Raggazzini, J., and G. Franklin, *Sampled-Data Control Systems*, McGraw-Hill Book Company, Inc., New York, 1958.

Ruchkin, D., "Linear Reconstruction of Quantized and Sampled Signals," *IRE Trans. Commun. Systems*, 350–355 (December 1961).

Schweppe, F., class notes from Course 6.603, Massachusetts Institute of Technology, Cambridge, Massachusetts, Spring 1967.

Smith, H., *Approximate Analysis of Randomly Excited Nonlinear Controls*, MIT Press, Cambridge, Massachusetts, 1966.

Steiglitz, K., "Transmission of an Analog Signal Over a Fixed Bit-Rate Channel," *IEEE Trans. Inform. Theory*, **IT-12**, 469–474 (1966).

Striebel, C., "Sufficient Statistics in the Optimum Control of Stochastic Systems," *J. Math. Anal. Appl.*, **12**, 576–592 (1965).

Swerling, P., "First Order Error Propagation in a Stagewise Smoothing Procedure for Satellite Observations," *J. Astronautical Sci.*, **6**, no. 3, 46–52 (1959).

Vander Velde, W., class notes for Course 16.37, Massachusetts Institute of Technology, Cambridge, Massachusetts, Fall 1967.

Widrow, B., "Statistical Analysis of Amplitude-Quantized Sampled-Data Systems," *Trans. Am. Inst. Elec. Engrs.*, Pt. II (Appl. Ind.), 555–568 (1960).

Wiener, N., *Extrapolation, Interpolation, and Smoothing of Stationary Time Series*, MIT Press, Cambridge, Massachusetts, 1966.

Wilkinson, J. H., *The Algebraic Eigenvalue Problem*, Clarendon Press, Oxford, 1965.

Wonham, W. M., "Some Applications of Stochastic Differential Equations to Nonlinear Filtering," *J. Soc. Ind. Appl. Math., Ser. A*, **2**, 347–369 (1964).

INDEX